BK 621.38 B985F
FUTUREWORK : WHERE TO FIND TOMORROWS HIGH-TECH JOBS TODAY /BUTLER, DI
1 C1984 16.95 FV

3000 646720 30016
St. Louis Community College

S0-BST-380

621.38 B985f
BUTLER
FUTUREWORK : WHERE TO FIND TOMORROW'S HIGH-TECH JOBS TODAY 16.95
FV

St. Louis Community College

Library

5801 Wilson Avenue
St. Louis, Missouri 63110

FUTUREWORK

FUTUREWORK

WHERE TO FIND TOMORROW'S HIGH-TECH JOBS TODAY

DIANE BUTLER

Holt, Rinehart and Winston
New York

Copyright © 1984 by Diane Butler
All rights reserved, including the right to reproduce
this book or portions thereof in any form.
Published by Holt, Rinehart and Winston,
383 Madison Avenue, New York, New York 10017.
Published simultaneously in Canada by Holt, Rinehart
and Winston of Canada, Limited.

Library of Congress Cataloging in Publication Data
Butler, Diane.
Futurework : where to find tomorrow's high-tech
jobs today.
Includes index.
1. Technology—Vocational guidance. I. Title.
T49.5.B87 1984 621.38'023'73 83-10797
ISBN Hardbound: 0-03-061984-X
ISBN Paperback: 0-03-064098-9

First Edition

Designer: Robert Bull
Printed in the United States of America
10 9 8 7 6 5 4 3 2 1

ISBN 0-03-061984-X HARDBOUND
ISBN 0-03-064098-9 PAPERBACK

To my son, Daniel, age four
—the future is yours.

CONTENTS

PREFACE

I used to think that people became authors because they wanted to, but I know differently now. Sometimes innocent bystanders get involved in that mind-bending process known as writing a book because they are driven to it. That's how it happened to me.

We often hear that hard times and high unemployment cause horrendous personal suffering that no statistical study could ever measure. While that is undoubtedly true, it was the faces of my students that brought the reality of this statement home to me. Hard times can crush the spirits of even those who have not yet entered the job market.

The college years are a special time of life. The young anticipate the future that is opening up around them with great excitement and the highest of expectations. In all the long years ahead, they will never again look forward quite so eagerly or quite so optimistically to what's coming next. It is a beautiful and unrepeatable time of life. For teachers, watching and guiding students at this moment of their lives are not only our prime duties but also our greatest reward.

That moment is also, as I discovered a few years ago, a time when students are particularly vulnerable, for it is easy to destroy the newly acquired confidence of the young. The word had gone out and all the students had heard it: There were no jobs out there, no one wanted them. Such a message is not easy for anyone to receive, but it can be doubly devastating to young people who are making decisions which will affect them for the rest of their lives. The enthusiasm of my students, which I had always taken for granted, disappeared. Teaching hopeless and cynical students was a far less joyful experience.

I didn't know quite what to believe or to say to my students. We must after all, I thought, be realistic. If the times are bad, even the young must know. Fostering foolish and unfounded hopes in an illusory future would be, I knew, far worse than telling the truth, no matter how depressing the reality might be. I knew little of the actual situation in the workplace but neither, I suspected, did my students. From the beginning I felt there was something wrong with their negative view of the future. It hardly seemed possible that the richest continent on earth, the land born of optimism, had given up the ghost so suddenly and so meekly. Was the news from the workplace really so bad? For me, finding out the truth about my students' fate, so intimately linked with my own, started out as a simple search, but quickly evolved into a kind of quest.

"High tech" was a catch phrase that I, like everyone else in America, had heard and I naively took it to mean computers. I had also heard that computers were destroying jobs, not creating them. But were they? At first I looked in books and libraries, as I had been trained to do. I soon saw that the answers to my questions were not in books or, indeed, in any one place. Then I realized that I had been overlooking significant data. Suddenly I saw clues everywhere: magazines and newspapers reported provocative tidbits, television regularly blipped out bits of information. My greatest source of information, however, was the people I sought out who were actually part of the working world. Before long, I found myself questioning (*grilling* is perhaps a better word) almost everyone I met. What is really going on out there?

Gradually I collected the evidence. No one person or printed source provided an overview, but I started to make connections between what I had seen or read and what people were telling me. A picture suddenly emerged, for I had stumbled into the new world of information technology that is now quietly shaping a new America.

The new world of telematics is already here, it is already well established. And it is fascinating. We, the general public, are unaware of its dimensions because we hear about it in so disjointed a way. Lacking an overview, we are hardly in a position to assess its significance for work or anything else. As a result of our ignorance, the world of telematics appears chaotic, unorganized, and tentative. It is none of these things. Video games, semiconductors, home computers, space shots, and even television are each only part of a far larger whole. The whole is made possible by technology and hardware, but that whole is much greater than a few new machines.

The whole made possible by telematics holds infinite promise, not only for us, but also for our children. They will live in another world. Creating, extending, exploiting, and maintaining that world is our job. It will take well into the next century. We are to make a new kind of America. Like our pioneer ancestors, we have gargantuan tasks ahead of us, technologist and non-technologist alike.

I now know that my students were wrong to feel hopeless about finding meaningful work in the future. They misinterpreted the message from the marketplace. The old world of the Industrial Revolution that we inherited from the past is clearly in upheaval. This chaotic period is certainly creating serious difficulties. But this period of chaos is a fertile one, for the chaos is caused by the coming of a tremendous new age.

In this book, I have tried to show how the world of telematics works. Once you know how that world operates, you will see, as I did, where the jobs are. My hope is that I have done even scant justice to the stimulating, exciting, challenging, and rewarding world of futurework.

DIANE BUTLER
Montréal

ACKNOWLEDGMENTS

Without the collaboration of many, many people, I could never have completed this book. To all those who generously gave their time, technical expertise, and sympathetic attention, I offer grateful thanks. Any errors still remaining are, of course, entirely my own responsibility. The names of many of those who helped me are scattered throughout these pages, but there are others whose contribution I particularly want to mention. Helene and Larry Hoffman, my associates and friends, offered unflagging optimism and enthusiastic support from the beginning to the end of what often seemed to me an impossible and interminable project. Carol Wilson researched and wrote the first draft of Chapter 5, "Television Comes of Age." Tom Ložar made substantive critical comments on the whole and considerably lightened the load of the final days with his humor and general craziness. Francis McInerney generously took the time to express his belief in my abilities and so gave me the confidence I sorely needed to scale those mountains of material. Coca Utreras, dactylógrapha extraordinaria, me ayudó más que solo palabras pueden decir.

FUTUREWORK

CHAPTER ONE

THE FUTURE STARTS NOW

Today's job market has become increasingly unpredictable and unstable. It is more and more difficult to make those crucial career decisions because the future of so many jobs looks so uncertain. In the past ten years, the previously unthinkable has happened in North America—massive unemployment among the young and not-so-young, huge layoffs in vital industries, retrenchment in many "growth" sectors, forced early retirement. There no longer seem to be any "safe" careers—teachers are unemployed in growing numbers, MBAs and lawyers are a glut on the market, and auto workers see jobs disappearing in their industry.

For the first time in fifty years, North Americans are reexperiencing the shock their parents and grandparents felt at the sudden arrival of the Great Depression. The older generation assures us there is no real comparison between soup kitchens and unemployment checks, but that is small comfort to those without work. We now know that our economic difficulties are a worldwide phenomenon; we are told the situation is likely to last indefinitely. This knowledge is not likely to provide much encouragement to job seekers. Facing economic decline on all sides, everyone looking for work these days wants to get in on the ground floor of an industry with a future. In a world of escalating employment insecurity, the job seeker must have the answers to two key questions: What are the new growth industries, and what does it take to enter them? Answering those questions is the purpose of this book.

THE NEW BREED OF WORKER

Work for pay is a central aspect of life. It has come to be the primary way that most adults, regardless of sex or age, identify themselves. Unemployment, therefore, has very real psychological, as well as material, effects. Mass unemployment compounds these effects, making it one of the gravest issues the nation faces today. Working itself is a basic human need; doing satisfying work is next in order of importance. When people must work below their abilities, underemployment is another, more subtle, yet also potentially catastrophic, effect of economic crisis.

Most people need to feel satisfied with work in order to maintain a sense of psychological well-being. This is one of the major conclusions of a famous study, *Work in America* (1972), published under the auspices of the Department of Health, Education and Welfare. Doing satisfying work confers a sense of identity and self-respect and brings order to the life of the individual. When the need for satisfying work is not met, the study shows, the consequences can be serious: boredom, absenteeism, a decline in physical and mental health, increased alcohol and drug abuse, and certain forms of adult delinquency. The individual certainly loses. When worker dissatisfaction is widespread, productivity falls and the nation as a whole suffers.

Despite the threat of unemployment, today's worker is demanding more interesting, creative, and flexible work. Since the 1960's, the North American attitude toward work has been undergoing a profound transformation. The aspirations of the work force have risen. Part of the reason for the change is the radical restructuring of the labor force. The work force is now younger than ever before. It is more educated; white-collar workers became half the labor force in 1978. Women represent 41 percent of the work force; the representation of minorities is now more evident. The labor force is more heterogeneous than at any time in the past. There is greater social and geographic mobility among the young, the educated, and minorities. These people are an entirely new breed of worker.

Frustrating the aspirations of the new breed provokes problems, as employers have already discovered. There is a growing tendency among workers to achieve "flexibility" through absenteeism and lowered productivity. In fact, the rate of productivity has declined markedly in the past ten years. It will continue to do so until the workplace has evolved sufficiently to satisfy the aspirations of the educated, mobile, and heterogeneous new workers.

This demanding new breed is the present and future reality of the labor force. Soon they must face another challenge. Beginning in the 1980's, all "baby boom" children will have entered the job market. Throughout the decade, there will be an enormous increase in the number of prime age (twenty-five to forty-four) workers. Prime age workers typically are competing for promotions and supervisory positions. This is the time of life when workers rise to the top of their professions and begin to enjoy the fruits of the long years of preparation. But the imminent remarkable increase in the size of this age group means fierce competition for jobs and substantial career disappointments for many people. In fact, those in the twenty-five to forty-four age category who have already suffered from being born into the largest generation in history may receive especially low relative incomes for their entire lives. How will middle-aged baby boomers, born to the rising expectations of the 1950's and 1960's, cope with the growing frustration of having no place to go? For people in this age group career planning today could well make the difference between a life of frustrating competition amid a sea of bodies and a smooth ride to the top.

Finding the right job is often difficult for the many young people about to make career decisions or to enter the labor market for the first time. It can also be an intimidating process for those presently holding jobs and looking for a change. These groups seek more than a mere job. Today's workers are no longer satisfied to plug away, day after day and year after year, in the same old boring way just to make money. They are searching for jobs that are challenging, dynamic, creative, and above all, right for the individual. Gone are the days of the thirty- or forty-year man. Frequent job changes have become commonplace. More and more people are looking for work that can satisfy their aspirations.

The uncertain job market and the new breed of workers' rising expectations make choosing a career or changing careers both more problematic and more important than ever before. Until now, there has been no book to help the millions of this new breed of workers find those satisfying jobs with a future. The technology of tomorrow explained in this book will provide many fascinating jobs. But understanding new technology will only partially prepare the job seeker for tomorrow's reality. Work in the near future will undergo a myriad of changes too vast to discuss adequately here—changes, for example, that will affect the very nature of the workplace itself.

WORKPLACE OF THE FUTURE

A six-month work year. Long leaves of absence to do socially useful work. Permanent part-time employees. A four-day week and handpicked holidays. Half-day work and half-day study. All these alternatives and more will be common in the workplace of the nineties.

In attempting to satisfy the expectations of the new breed of educated worker, employers are experimenting with a great many employment options. Such innovations, called variously the human resources movement or organizational development, cut across all strata of the labor force. Blue-collar and white-collar worker, professional and factory hand alike can benefit from the new approaches. The old ways simply do not work anymore. Employers have found that putting effort into restructuring the workplace in a more human way is merely sound business practice. Industry reaps the benefits of higher productivity, improved labor relations, fewer strikes, and improved company loyalty.

Change, flexibility, and innovation are the hallmarks of the movement. There is no longer a standard way of doing things, as there was in the past. The labor force is now too diversified for that. What suits one company's particular conditions may not suit another's. What appeals to one employee falls flat with another. Variety is the key to the workplace of the nineties. This quiet evolution is expected to take until the end of the century to complete, but its major impact will be felt in the 1980's and 1990's.

The new breed of worker wants not only more interesting jobs but also more control over jobs and less rigid divisions among education, work, and leisure. Traditional time management is changing as the lines separating work, study, vacation, and retirement shift and fade. Flextime, or gliding time, which allows workers to put in their eight hours anytime between 6 A.M. and 6 P.M., is an alternative in some companies. The shift with a variable number of hours—six one day and ten the next—works for others. Or an employee may pick from a variety of schedules—for a month at a time. For those over fifty-five, a six-month year that gradually phases out the older and phases in the younger worker may seem right. McDonald's pioneered the four-and-a-half-day week; in the nineties the four-day week will be commonplace.

Part-time workers now account for 21 percent of the labor force. Many of these people do not want full-time work, and a variety of options have been opened to accommodate them. Paired, or partnership, employment, say, two elementary-school teachers sharing the day, is one. Another is shared work in which the day is not split, but the work is, as, for example, in welfare casework. Or split-level work, where two people perform separate functions, such as composing and typing letters. Or split-location work. The possibilities are endless.

The mounting drive for variety has great implications, for it means sweeping away lingering relics of the past now entrenched in the workplace of the present. The introduction of assembly-line techniques was one of the major progressive innovations of the past. Because of the standardization of parts and the division of labor, productivity soared beyond our wildest dreams. What worked in the factory was applied to the office. The goals of the large modern office, characterized by row upon row of desks or compartmentalized cubicles, are productivity and speed, accomplished through job fragmentation and standardization of procedure. The office has become a replica of the factory, even though its product is information rather than a durable good. Today, fixed rates and standards in the workplace, legacies of the assembly line, are the preference of unions, governments, and large companies. The norm is standard from coast to coast: one rate of pay, one length of workweek, one set of job descriptions, one rate of output.

We have succeeded in applying assembly-line values to many aspects of modern life, but we have paid a very high price. Robot-like lives, apathy, depression, disillusionment, and senseless thrill-seeking are all to some degree or another results of forcing the mass of people into predetermined, standardized slots. Workers with Depression-era mentalities sought security and stability above everything else; acting like a machine seemed a small price to pay. Increasing affluence and education have changed all that.

The youth culture of the sixties and seventies produced a more tolerant attitude toward individual differences, a greater acceptance of variety in personal behavior and appearance. At the same time, minorities and women became increasingly more vocal in their demands to be accommodated in the mainstream of our society. Ethnic dissimilarities are now emphasized, rather than suppressed. All this points to an increasing heterogeneity in our culture. Conformity and standardization are no longer the norm and will be less so in the future. The traditional assembly line is a back number. The norm of the future is to have no norm. The technology of the future could not come into full flower in a homogeneous environment.

JOB SKILLS: GETTING DOWN TO BASICS

The average employee holds twelve different positions in a lifetime. This number is likely to increase in the future. Few people today work throughout their lives at the same job they were trained for. This trend, too, is likely to continue. Although specialization may have been an appropriate response to yesterday's job market, the key to success in the future is a more generalized approach.

As yet, we do not know how to restructure our educational system to meet the changing needs of tomorrow. We do, however, know, at least in outline, what job skills will be valuable in the near future. The successful jobholder of the future will need two types of basic skills. The first is technical. The second is the ability to communicate effectively. These two skills will be more interrelated than in the past.

In the recent past, familiarity with computers was necessary only for certain technical positions, but the "information explosion" of the eighties will change that. Each and every one of us will be required to have some knowledge of computers. Computerphobics will hate the future. More important, they will increasingly be unable to function there. All of us will have to be able to handle an alphanumeric keyboard and to perform simple tasks on a computer terminal. Of course, for those with career aspirations in the emerging technologies, a much wider range of technical skills will be necessary. And those now in the work force who hope to change jobs should remember that even high school graduates of the 1980's will be conversant with at least one computer language.

As computer technology permeates every aspect of life, the workplace will, surprisingly, become more human. When machines perform routinized tasks, people are freed to deal with ideas and with each other. This necessarily entails a greater emphasis on communication.

Securing a job in the nineties will indisputably require communication skills. Success in the business world depends upon expressing ideas to others. Even engineers and others in technical fields must be able to do so. The most rigorous experimentation is worthless if it cannot be presented to others in a comprehensible form. In the nineties, those with the mental agility to cope with the rising tide of information, to make decisions, and to deal effectively with others will be premium goods.

Traditionally people communicate on the job by listening, speaking, and writing. This will not have changed by the nineties, but by then workers will also need skill with the electronic media in order to communicate effectively. Interpersonal communication is important in today's workplace. Both public communication and person-to-person interchange will be augmented and in some cases replaced by audio and video productions. Those who do not know how to express themselves through these media will be at a distinct disadvantage.

FUTUREWORK: HOW WE KNOW

No one can know the future. No one can predict with absolute accuracy what jobs will exist ten, or even five years from now. Fortunately, however, serious research into the future has become increasingly common. Forecasting probable jobs, though difficult, is not now impossible. Futurologists attempt to make forecasting the future or, more precisely, alternate futures as accurate as possible through the use of techniques of varying powers and potential applications. At last report, there were no fewer than 150 techniques available, although only seven or eight are commonly used.

The findings of such forecasters are extremely useful, but their limitations must be understood. Futures research is not prophecy. The future now, as always, is not knowable in advance. The growth, present and future, of certain areas of the economy results from a concurrence of factors which are often elusive and complicated. These factors include social innovations, economic events, changes in the family structure, demographic trends, and political decisions. Some future developments are more predictable than others. In general, economic opportunities resulting from certain types of technological innovation are more predictable than, say, the social consequences of a new invention.

What happens in the present makes the future. Thus, forecasting for the next five or ten years is more likely to be accurate than speculations about the year 2050. It is also extremely important to realize that present expectations about the future can, to some extent, make that future happen.

Projections of futures researchers, extensive interviews with experts in the field, and statistical evidence tell us that several areas of the economy will experience growth or radical change by the 1990's. The steel and automobile industries, for example, and, indeed, most traditional manufacturing will change enormously in the next ten years. The first shock to the auto industry is even now painfully apparent. We already know what gearing this industry to a new reality will require: massive robotization and more reliance on computer-aided design and manufacture (CAD/CAM). New technology is also creating many now-emerging industries. We recognize, for example, that biotechnology is an industry of the future, which will require years, perhaps a decade or more, of basic research; the promise of this technology may not become the reality of a viable industry until the turn of the new century.

It is possible to predict with perfect accuracy the important technologies of the future because they are now in the works. Gigantic new industries do not spring up from one day to the next. Even the most exciting new industries of the past—the automobile, aviation, moving picture, and camera industries, to name just a few—needed a period of at least five to ten years for research and development. Breakthroughs in technology are not immediately available on the market. First the market must be carefully analyzed. The needs of the marketplace must cry out for the development of a new technology, or the need for a new technology must be created in the marketplace. This process does not change.

Tomorrow's industries are here today. And we know which they are. *Telematics has already emerged as the mega-industry of the century.* The term telematics refers to an astonishing concatenation of subindustries, some familiar and well established, others barely out of the research stage. This greatest of the coming new industries subsumes what were previously separate areas of endeavor: the communications, information, and broadcasting industries. In the past, communications was a clear-cut industry that included the telephone, the telegraph, and Telex. The information industry, data processing, is a result of the computer revolution of the past thirty years. Gradually, however, communications and data processing have become increasingly interlinked. In fact, though it is not yet widely understood, these two previously separate industries have actually merged into one. There is no longer any real distinction between a computer terminal used for communication and one used as an input/output device for a computer system. The markets of the two industries are now identical, served by both telecommunications and data processing companies. Digital technology, formerly limited to computers, is also making changes in the electronic media, especially television. The telecommunications sector, the information sector, and broadcasting are merging, because of their common use of digital technology, into one mega-industry. Called variously compunications or telematics, this massive industry has enormous scope. Its impact on the job market of the near future will be nothing short of stunning.

HOW TO READ THIS BOOK

The world of technology is changing so rapidly, it has become almost impossible to keep up to date. Every day, we read about a new technological breakthrough in the newspapers or see it on television. How do all these breakthroughs fit together? What do they mean for you and your future? We all know more change is coming. To prepare ourselves, we must understand, at the very least, what has already taken place.

This book is an introduction to the world of the very near future. That world will be different, but it is evolving from the world we live in now. Even the seemingly arcane telematics industry is comprehensible; its various subindustries fit together in an organized way. This book lays bare that organization and shows how the ramifications of this industry only begin with technology. The employment possibilities generated by this emerging industry go far beyond the purely technical occupations we would expect to find. In providing a guide to future work, listing titles of jobs that don't even exist yet would obviously be of little use. *Futurework* cannot be like other job books. *Futurework* provides a detailed outline of the coming new world of technology and gives many examples of work to be done there. Seeing the direction the technology is taking makes it easier to discover others.

This book explains the technologies, translates the technogab, and eases our entry into the world of tomorrow. In doing so it teaches a method, a way of looking for openings. The future starts now: It comes about as a result of what we are doing today. If we know where to look, we can figure out what will be going on tomorrow. If you get in on the ground floor, you will be among those who make the future happen.

The jobs of today may be in the want ads, but the jobs of tomorrow are on the front pages and on the drawing boards. This book is not the want ads. It is the news. It is a trip through the modern world. As we go through it, you will begin to notice many Help Wanted signs in the windows. After finishing this book, you will be surprised at how many you've seen. At least half of them aren't in these pages at all. Because you've learned a method, you will have figured them out for yourself.

CHAPTER TWO

THE KEY TO HIGH TECH

SEMICONDUCTORS

Many people have never heard of the semiconductor, yet in the short span of ten years the semiconductor industry has become the shining star of American industry and the focal point of an entire planetary system of satellite businesses crucial to present and future national prosperity. Through the creation and development of semiconductor products, such as the transistor, the integrated circuit (IC), and, most exciting of all, the fabulous microchip, this industry has almost single-handedly brought into existence an impressive agglomeration of other new industries, now collectively labeled high tech. Without an understanding of the undeniably complex semiconductor products and their technical significance, it is impossible to grasp the potential of high tech to change our lives. Semiconductors are the key to technological innovation for the 1980's, the 1990's, and probably well beyond.

By spawning microprocessors, the legendary microchips, semiconductor technology has generated an avalanche of innovative products, from calculators and digital watches to Space Invaders and industrial robots. Semiconductors are the basis of today's $100 billion world electronics industry, whose record growth rates seem to know no bounds. Perhaps even more significantly, semiconductor products are transforming traditional technologies, from automobiles to appliances to airplanes. The familiar business office wears a new face because microchip technology has altered the typewriter, the telephone, and the file cabinet. Heavy industry is not immune. Robots, sophisticated process-control equipment, computer-aided design and manufacture, electrically controlled machine tools, all made possible by a wafer no bigger than a cornflake, liberate humans from most of the dirty, dangerous, and monotonous occupations of the workplace. The traditional professions are next: education, law, medicine, and accounting have only just begun to feel the transforming touch of the tiny chip. In ten years, microchip technology has affected almost every aspect of modern life. Whether we realize it or not, behind that stove, clock, or camera, behind the family car, television, or smoke detector, lies a microchip brain. This tiny piece of silicon has rocked the marketplace, creating multibillion-dollar industries and in the process spawning hundreds of thousands of new jobs. Ten years ago, most of these jobs did not exist. By 1990 the pioneering accomplishments of the first ten years will seem like child's play.

In fact, the semiconductor industry gave birth to the much vaunted Information Revolution, which, even now, in its unruly and unpredictable infancy, is completely changing our society and our way of life, much as the Industrial Revolution altered the quiet rural lives of our ancestors. Then mechanical innovations of all types, from the train to the typewriter, eventually coupled with electricity, were the watershed technologies of the day, bringing with them monumental benefits. For better or worse, as the Industrial Revolution raised our standard of living to unprecedented heights, it totally obliterated the old ways. Without much fanfare, our familiar electromechanical world, end result of the Industrial Revolution, is now sounding its death rattle. Taking its place is a new world, promising benefits just as enormous, but which will, at the same time, also alter our lives to an almost unimaginable degree. The new world is possible thanks in large part to the semiconductor industry, quietly flourishing in almost complete obscurity.

WHAT ARE SEMICONDUCTORS?

Semiconductor products, like the computer itself, are almost exclusively American inventions. To understand how semiconductors can be so important, first we must know a little bit about the computer and how its developmental difficulties led to the invention of semiconductors.

In essence, a computer is a vast collection of switches. All information, whether words or numbers, fed into a digital computer is translated into a binary numbering system. The computer receives all data as binary digits. Each digit, called a "bit" for short, is either a *0* or a *1*. A *1*, meaning "on," represents a closed switch and a *0*, meaning "off," represents an open switch. To illustrate long trains of combinations of these two numbers, a computer needs many switches. For a computer to have any utility, the switches must be mind-bendingly fast. When the first true digital computer, ENIAC, appeared in 1946, the fastest switches were vacuum tubes. ENIAC's circuits used 18,500 vacuum tubes for automatic switching.

Using vacuum tubes, though, posed many difficulties and curtailed the future development of the computer. Vacuum tubes had an average life-span of twenty hours. With thousands of the glowing hot tubes in a single machine, some computers shut down every seven to twelve minutes. The size and power of planned second-generation computers seemed severely limited by the shortcomings of the vacuum tube. During the late 1940's, Bell Labs, then as now one of the largest research and development centers in the world, began searching for a substitute for the troublesome vacuum tube, used also in telephone switching equipment. By 1953, two inventions, both called "transistors," had surfaced; Bell's William Shockley was a corecipient of the Nobel prize for his work in their development.

The transistor, a thumbnail-sized sandwich of solid material, ushered in the era of "solid state" electronic devices. Transistors are microscopic crystals made from a semiconductor material; in contrast to vacuum tubes, they are small, use little power, and generate little heat. Transistors not only boosted the consumer electronics industry, making possible portable radios and hearing aids as well as higher-powered audio components, but also transformed the fledgling computer industry. At the beginning of the 1950's, only a handful of room-sized computers were in existence; generally only the government had the funds and space needed to make their use economically feasible. Silicon technology provided the basis for three generations of increasingly compact, reliable, and cheap computers over the next twenty years. By the 1960's, the now economic computer had begun to proliferate.

The transistor is a relatively simple device. All electricity flows in currents. Some metals, such as copper, aluminum, iron, and even silver and gold, are good conductors; electricity flows through such metals easily. When an electric current flows through a conductor, there is no preferred direction of flow; power can be tapped from any place. Certain elements, called semimetals, or metalloids, because they have some of the properties of metals, especially silicon, are semiconductors; electricity flows through them, but only in one preferred direction, while weakly or not at all in the opposite direction. So a semiconductor, as a one-way-flow device, is most useful in operating a switch where current must always flow in the same direction. The vacuum tube is also a one-way-flow device, but its filaments are made of conducting material. The silicon-based transistor could easily replace the much larger, clumsier, and hotter tube. Since silicon, which is essentially sand, makes up about one quarter of the earth's crust, semiconductor devices could be extraordinarily cheap if produced in quantity.

Shockley was aware of the potential of his invention and immediately left Bell to establish his own lab for commercial production. This laboratory, located along the south shore of San Francisco Bay in the Santa Clara Valley, became the nucleus of the nascent semiconductor industry. Assisted by capital from Fairchild Camera and Instrumentation Company, Shockley attracted numerous brilliant designers from nearby Stanford University to his lab. Gradually his recruits struck out on their own to found new companies, such as Intel, Signetics, and National Semiconductor. Called the Fairchildren because of their relationship with the parent company, at least fifty of these firms are still in the valley today. Now, Silicon Valley, as it was soon dubbed, is home to some 500 electronics firms and is still the center of the semiconductor and world electronics industries.

Around 1960, the integrated circuit (IC), now the dominant circuit module in practically all electronics, first appeared. By means of photolithographic printing or etching, ICs combine two or more individual transistors on tiny slices, or chips, of semiconductor material, usually silicon. By adding microminiaturized circuit elements to the same chip, engineers were able to fabricate a practical device that rapidly became more sophisticated as more and more switching circuits could be packed onto a single chip. Eventually these silicon chips contained the equivalent of thousands of transistors. IC chips found a variety of applications. In the generation of the IBM 360 computer, still the industry standard, ICs were responsible for a significant improvement in the performance/price ratio while decreasing the size of the machine and increasing its reliability. ICs also became common in such consumer products as dryers, dishwashers, and even toys. IC chips are simpler, more reliable, and cheaper than the electromechanical parts they replace.

While the cost of an IC has remained constant, the number of transistors packed onto the chip has increased nearly exponentially throughout its history, leading to dramatic reductions in cost relative to function. This sounds straightforward enough until one considers what it really means. In 1958, 10 electronic components could be placed on a single silicon chip. By 1970, 100 units could fit on the chip, a tenfold increase in slightly more than a decade. But then, by 1972, 1,000 units could be put on a single chip and, by 1974, 10,000 units. By 1982, well over 100,000 units could be fit on the same chip, with the 1,000,000 mark expected any day.

Such spectacular industry advances are absolutely unprecedented. To gain some perspective on the achievements of the electronics and semiconductor industries, consider this: Throughout the entire history of autos as a leading growth industry, car manufacturers made trivial design changes at three-year intervals. To approximate the performance of the electronics industry, Detroit would have had to increase gas mileage from 10 miles per gallon in 1958 to nearly 1,000,000 miles per gallon in 1985. Of course, in reality, even under intense pressure from foreign competitors, American car manufacturers boosted gas mileage from 10 to 30 or 40 miles per gallon during that period. Standards of achievement in high-tech industries leave others gaping.

BIRTH OF THE MIGHTY MICROCHIP

Until the early 1970's, the semiconductor industry's prime product was the miniaturized IC chip. While improvements in the IC chip represented tremendous progress, they are but a small part of the story of the electronic revolution. The truly earthshaking innovation was the development of the microprocessor, invented by Ted Hoff of Intel Corporation.

At the time, it seemed a negligible event: the November 1971 issue of *Electronic News* carried the first ad for a microprocessor, Intel's 4004. The microprocessor seemed a relatively innocuous invention because no one, including its designers at the then tiny California company placing the ad, had any idea of what its applications might be. Although its development was often foretold during the 1960's, when the awaited event actually occurred, nobody really understood its significance. Intel staffers thought it had a few dozen potential applications around the home, none of which has proved viable. For the microprocessor does not represent an increase in magnitude like the IC chip but, rather, a change in *kind*. This change in kind of circuit integration enabled the electronic revolution to gather speed and finally to become accessible to the general public through cheap home computers.

Like the IC, the microprocessor is actually a series of razor-thin silicon chips, etched with complicated circuit-designs and then bonded together. But while each IC chip is custom-designed for use in a highly specialized environment, the microprocessor is general purpose. The microprocessor is a central processing chip that contains a simple core of logic. As a central processor, the tiny microprocessor has the abilities of a big computer—it can perform operations on data according to a series of instructions and can perform logic and "decision" functions. The core of logic serves as a brain when the microprocessor is hooked to memory chips to form larger systems. One or two of these chips could run a complete computer.

In short order, as the light dawned on its inventors, the humble microprocessor, now better known as the fabulous computer-on-a-chip, sent shock waves through many an industry. For the microchip can provide the intelligence to run an incredible variety of machines. The watch industry was one of the first to succumb to microchip technology. At first, Swiss and German manufacturers looked on in tolerant amusement when electronic watches with microchip "brains" instead of electro mechanical works appeared. The smiles vanished as American and later Japanese upstarts took over half the market and all but destroyed the European watch industry. Calculators were next. Today, one looks in vain for an electromechanical version. Every schoolchild packs a plethora of calculating power in a ten-dollar hand-held model. Other consumer products from stereos to video games will never be the same, thanks to the microchip.

Important as the microchip has been in its industrial and consumer products applications, nowhere has its impact been felt more than in the computer industry. Even the mighty giants of the computer industry, the likes of IBM and DEC, have been shaken by the awesome powers of the fragile chip—although contrary to popular belief, established computer companies had nothing whatsoever to do with the development of the microprocessor, as we have seen. Microprocessors are the "brains" of the microcomputer. Microcomputers are not scaled down versions of the big machines; they are a wholly different species. The microcomputer industry, born in 1975 as a consequence of the microprocessor, is a child of the semiconductor industry.

In the years since Intel put the first microprocessor on the market, numerous improvements have been made. The most important increase in complexity involves putting more functions onto a single chip in order to reduce the amount of external circuitry needed to make it usable. The earliest primitive microprocessors required a large amount of external circuitry. Typically each chip needed thirty to fifty additional circuits before it could be used. Now much external circuitry has been eliminated, and as a result the price of a chip has fallen dramatically.

The microchip is so special that it is worth examining in detail. The microprocessor used in the popular Apple II home computer, for example, looks something like an aerial photograph of a railroad switching yard when viewed under extreme magnification. Each of the tiny tracks acts as a kind of branch line for electrical signals passing through the central processor. The chip's surface has three main areas. A control unit makes sure that instructions are executed in the correct order; it can be said to regulate the switching of tracks. The arithmetic-logic unit functions as the switching yard to actually execute essential programming commands, such as addition, subtraction, multiplication, division, and logical comparisons. The third part functions as a temporary memory storage device that allows the microprocessor to remember what it is doing at the moment. Another kind of chip, called a memory chip, acts as memory storage for actual information or data. When hooked together, these two kinds of chip form a computer. To the naked eye, they may look identical, but under extreme magnification their differences are patent. Instead of the complex branching tracks of the microprocessor, the memory chip has an orderly gridiron pattern, like a city street map.

Semiconductors are not only an almost totally indigenous creation, but one all Americans can be proud of. Today, in addition to Silicon Valley in California, minicenters of the semiconductor and electronics industry have grown up all across the country. Silicon Alley, running along Route 128 near metropolitan Boston, is one location; another is the belt around Washington, D.C. Silicon Prairie refers to the Dallas-Fort Worth area, but outposts of such companies as Texas Instruments and Mostek extend its boundaries to include San Antonio and Austin. The Research Triangle in North Carolina is fast becoming yet another such center. In Canada the region north of the nation's capital, Ottawa, alias Silicon Valley North, harbors a concentration of such firms as Mitel, Gandalf, and Nabu.

Although it is too early to tell if the minicenters will rival the success of the original Silicon Valley, studies have shown that starting a high-tech company there is as near a sure thing as the free-market system offers. The average manufacturing company stands a 75 percent chance of surviving its first two years; Silicon Valley companies stand a 95 percent chance of surviving their first six years. Companies in the valley are this successful because of the complex network of universities, support companies, consultants, scientists, managers, and investors developed over the past thirty years to cradle new firms until they are ready to stand on their own.

THE SEMICONDUCTOR INDUSTRY: JOBS

ROBOTS OUST WORKERS; WILL THE CHIP GET YOUR JOB?; COMPUTERS TAKE OVER—HUMANS REDUNDANT. Do these headlines sound familiar? They should: similar ones appear in local newspapers every week. Doomsayers, and they are legion, love to tell us all how the increasing technologization of our society is going to do us all out of jobs in the very near future. Today, the refrain seems to be: If automation didn't get you, then the computer will. Lately it has also become fashionable to warn doctors, professors, accountants, and lawyers (add any other high-prestige profession you can think of) that their jobs will soon be on the line as well. Are the doomsayers right? Does technology destroy jobs? One simple answer is, yes, most new technology does destroy jobs. Think about it. How many blacksmiths or carriage- or harness-makers do you know? These were once common occupations, wiped out by the invention of the automobile. Then, too, Blanche, the nosy telephone operator, is long gone, and Mac, the iceman, comes no more, both victims of technological advances. The demise of the blacksmith and the replacement of the telephone operator with direct dialing make us realize what the doomsayers never seem to learn from the past: *Every new technology creates jobs that never existed before, often with stunning speed.*

Springing literally from the dust of the earth, the semiconductor industry is a splendid example of a brand new technology which, in just a few short years, has created thousands of new jobs. Producing first transistors and then integrated circuits, the industry was already exceedingly respectable when, around 1970, it pioneered the development of the microprocessor. And then its fortunes skyrocketed. Semiconductors became *the* growth industry of the 1970's. As more and more applications for its chip turned up, the industry rushed to develop still more powerful products. The number of functions per chip increased so dramatically that during the industry's entire history each year's production has equaled the sum total of all previous years' output. At the same time, costs plummet: today an entire computer-on-a-chip costs about as much as a single transistor did in 1960, and every year the price of the chip drops 30 percent. The advantages of semiconductor products are so great that the world demand for them quadrupled during the 1970's; it will quintuple during the 1980's. In fact, a major problem during the recent economic slump has been producing enough products to meet the escalating demand. During the past decade the market grew at an annual rate of 22 percent compounded and will continue to grow as spectacularly during the 1980's. According to the market research company Dataquest, worldwide sales of semiconductors will grow by nearly 15 percent in 1983, to well over 16 billion dollars. The American Semiconductor Industry Association estimates that American firms made 55 percent of the 14.5 billion dollars' worth of semiconductors sold throughout the world in 1982. Japanese companies had 35 percent of the business, the Europeans 10 percent. Predictions are that by 1990 annual worldwide industry sales will climb to something like $60 billion from the current $11 billion. The overall world electronics industry, based in large part on semiconductors, will see annual sales on the order of $400 billion by 1990, making it the largest business in the United States, bar none.

The development of innovative semiconductor products has had a direct effect on the labor force: the semiconductor industry itself employs about 250,000 people in the United States. In addition, a whole complex of associated and supporting industries have sprung up, from companies engaged in chip testing and measuring to integrated electronics firms. In 1979, electronics manufacturing was the direct employer of 1.4 million Americans. As a touchstone technology, semiconductors also feed the automobile, telecommunications, computer and data processing, entertainment, and military supply industries. The effect on the labor force of semiconductor technology in this feeder role can only be guessed at, but it is certainly staggering.

Unlike traditional industry, the new technology does not depend on expensive, non-renewable raw materials, vast pools of unskilled labor, or multi-acre factory sites. High tech is clean, safe, energy efficient, and highly profitable. For these reasons, local governments across the continent are making every effort to attract such an ideal employer to their regions. The single most important resource a locality must offer is access to abundant human brainpower, a highly skilled and trained work force. What the already established high-tech centers have in common is proximity to fine academic institutions. Thus, for example, as Silicon Valley has Stanford University, Boston offers the Massachusetts Institute of Technology and Northeastern and Harvard universities, while the University of North Carolina's Chapel Hill campus and Duke University serve the nearby Research Triangle.

The semiconductor industry, however, is currently experiencing a variety of difficulties. Ironically, some of its most significant problems stem from the very success of the industry in creating a rapidly expanding worldwide demand for its products. Adequately trained technical personnel seem to be in short supply. Recently industry spokesmen have called into question the ability of the United States to meet its escalating demand for engineers and other even more highly specialized personnel. Thus there is doubt as to whether America can maintain its hard-earned position as world leader in the production of semiconductors, the key to so many other industries crucial to the country's present and future economic prosperity. Currently the United States produces about 60,000 engineering graduates a year, about 6 percent of all those receiving undergraduate degrees. By contrast, in Japan 75,000 engineers graduate every year, representing 21 percent of all college graduates; the U.S.S.R. sees 300,000 new engineers a year, 35 percent of the students receiving first degrees. Clearly the United States must produce more engineers, but the question is how? Again, the semiconductor and related industries' very success appears to be creating the problem. Because positions within the industries are so highly paid, recruiting high-level staff is extremely difficult for the financially burdened universities.

The high-demand market is, however, a bonus to the qualified job seeker. Certain types of engineers are always in demand in the entire electronics industry, as we shall see again and again. The following types of engineers should have little difficulty finding employment in the semiconductor industry:

- electronic
- electrical
- mechanical
- materials scientist
- industrial management.

The semiconductor industry also has a strong need for

- mathematicians
- physicists
- computer scientists
- computer engineers
- information scientists

and in lower-level positions:

- electronics technicians
- draftspersons

and many others. Most of these specialties are obvious and self-explanatory, but several technical professions now gaining prominence in the semiconductor industry deserve mention for their oddity. Engineers-in-the-making may find them intriguing possibilities.

Ceramics engineering is the first. Working in a rather obscure profession, the ceramics engineer, until recently, was generally in the employ of glass manufacturers, such as Corning Glass Works or Coors Porcelain Company, producing a variety of familiar products: building materials, insulation, self-cleaning ovens, and light-sensitive glass, to name a few. Technological breakthroughs too complicated to explain here are today propelling the profession to the forefront in bioengineering, steelmaking, fiber optics, and aircraft and automobile design, as well as in semiconductor and microcircuit manufacture. In fact, various industries, including the semiconductor industry, now call for about 1,500 new ceramics engineers a year, while schools turn out a mere 500. Only eleven colleges grant degrees in ceramics engineering. The State University of New York's College of Ceramics in Alfred is the largest with a total of 400 undergraduates. Many engineering schools do, however, offer individual courses, usually within departments of materials and metallurgy.

Cryogenics, the science and technology of low temperatures, is the next surprise entry. In the past, cryobiology, investigating freezing techniques for sperm and blood banks, for organs used in transplantation, and for tissue cells, seemed to provide the most interesting avenues of research. The growth of the space industry with its need for materials able to withstand the cold of the cosmos stimulated developments in the field of cryogenics. The discovery that extreme cold transforms certain metals and gases into efficient, economical superconductors of electricity has great implications for the semiconductor and companion industries. In fact, among other industry-related applications, cryogenics may play a large part in creating the next generation of supercomputers. A new type of semiconductor, the Josephson Junction, functioning only in ultra-cold conditions, may replace silicon, producing operational speeds and efficiency previously thought impossible. The majority of present-day cryogenicists began life as physicists, chemists, computer scientists, or, most frequently, engineers, and then stumbled by chance onto this new area of research. Future cryogenicists will probably begin their studies in the same areas, but their entry into the field will be the result of more careful planning.

While qualified technical personnel should continue to have little difficulty locating jobs within the industry, business and financial professionals will begin to find new and ample opportunities there. The semiconductor industry is headed for major changes in the eighties. The industry, "wonder child" of the 1970's, has reached a critical stage in its development, for it is rapidly maturing. The semiconductor companies of the seventies typically originated as small concerns founded by engineers or designers striking out on their own. Success in those days hinged on technical innovation and little else; effective management techniques were not high on the list of priorities. Times have changed, however, and economic woes, endemic in the rest of North American manufacture, have not spared the semiconductor industry. Interest rates are high, and venture capital is scarce at a crucial time. Semiconductor firms are now middle-sized, and their increased financial complexity frequently requires more business expertise than many engineer-managers are able to muster. As a result, mergers have become common as the original engineer-owners are obliged to sell out. A most menacing development, however, is the recent rash of acquisitions by foreign buyers. Up to 60 percent of the take-overs in the industry are by foreign companies, who often make deals not only to get the product and market but principally to gain access to the technology itself. Semiconductors are as American as apple pie, but without firm financial organization, new capital formation, and expanded research and development programs, the industry or a large part of it may well be exported to other shores. The situation cries out for the skills of trained business professionals.

Considering the current and expected future shortfall of engineers, the industry's present practice of trying to turn engineers into managers may well change. Not only are the skills of experienced engineers needed for technical innovation, but many excellent engineers lack managerial skills, such as the ability to communicate, to delegate responsibility, to inspire, to lead, and to define jobs to others. The demand for professional managerial personnel within this industry will certainly continue to escalate. For many functions, however, the skills of that now rare bird, the trained engineer–successful manager, will be vital. Much of the sophisticated research necessary to the continued well-being of the semiconductor industry can only be translated into efficient production through astute managerial intervention. Realizing this, several universities are trying to fuse these two areas in an attempt to redress the serious management situation within high-tech industries.

Future engineers might do well to be on the lookout for these new programs combining engineering and management. Silicon Valley's University of Santa Clara, for example, sponsors the Early Bird Engineering Program, designed to give working engineers a chance to get a master's degree in engineering management. At the undergraduate level, Clarkson College of Technology in Potsdam, New York, offers courses combining management practices with engineering. Carnegie-Mellon University in Pittsburgh takes advantage of two of its strongest departments, the management school and the engineering program, to teach engineering to managers and vice versa. Boston's Northeastern University has just initiated a special graduate program in high technology which is open only to those who already have four to seven years of experience in industry. Opportunities for the 1990's look bright in the semiconductor industry.

CHAPTER THREE

TOYING WITH COMPUTERS

Computing as an industry is less than thirty years old. Few industries have sprung up so suddenly, spread so rapidly, or become such an integral part of our lives in such a short time. Already it is hard to imagine our world without computers. So far, the impact of computers, pervasive as it is, is most evident in the business world, though hardly confined there. Life without computers is almost unthinkable: schools, hospitals, banks, supermarkets, government services, transport, travel organizations are all computer-dependent. The ubiquitous credit card is a reality largely because of computers. The Post Office seems such a ponderous white elephant mainly because not all its services are computerized. Most of us take for granted the greater volume of information and higher level of sophisticated service made possible by computers.

To many of us, though, the changes in our surroundings brought about by computers may seem a mixed blessing. Computers have assigned us all numbers we can't even read as substitutes for our names. Delays, mistakes, and foul-ups of every imaginable type are now invariably ascribed to "computer error," rather than human fallibility. Presumably no real person makes mistakes anymore. At least no person can be found who is responsible for them. How many of us have waited helplessly while our bill, check, or order was "lost in the computer"? Products in grocery stores, including even books, carry bizarre markings, called the universal product code, which mean something to the all-knowing computer but nothing to ordinary human beings. These days bills and personal checks come accompanied with little squiggles that we all know are there for the computer to read. Who has not wondered: *Why send these squiggles to me? Why put them on my checks?* We all know there is information about each and every one of us located in secret data banks in faraway places. None of this is very reassuring. In fact, in many ways computers can be at the very least irritating and at the worst frightening. They tend to make us feel stupid and impotent. Big Brother may well be out there, and that awful presence seems perfectly exemplified by computers and their growing presence in our lives. The anxiety of the average person when a computer intrudes into his or her daily life is hardly surprising. Computers are so new and their use has spread so rapidly that a negative public reaction is almost inevitable. Most people over the age of fifteen never saw one in school or learned how one worked; many of us have never touched a computer. As a result, computers seem mysterious. Mystery is always frightening. For those untutored in the realities of the computer, the machine may seem omnipotent or endowed with superhuman qualities. Computers calculate and collate data so rapidly that they do truly appear to be magical.

Recently, however, the mighty computer has become friendlier. In the past few years video games and other electronic consumer products of all types are everywhere. The reaction of the public, previously doomed to feel intensely insecure about computers, is very interesting. Electronic games and toys are great hits with Americans, who, when given the chance, are obviously eager to banish fear and find out for themselves what computers are all about. Now we can all have a true "hands on" learning experience as we see, touch, and play with computers in our own homes. Every child crying for Pac-Man, Space Invaders, or Blockbuster turns his beleaguered parents into computer buffs, if only to protect their investment in an exceptionally expensive toy. Intellivision and Atari have shown millions that computers can be fun and, most of all, friendly. These games have inadvertently performed a tremendous public service by inoculating the public against the epidemic of computerphobia which has been spreading across the land for the past thirty years. If the average person can handle a joy stick, the tiny step to key pad or full-fledged computer terminal is an easy one. Since the immediate future promises more, not less, computer involvement in our lives, eliminating the psychological barriers is quite necessary. Computers cannot frighten us if we understand them.

In fact, it now appears that the American public's fear of the menacing computer is giving way to an out-and-out romance. Ten years ago only eccentric technocrats had computer terminals at home; they were treated with the same mixture of condescension and respect traditionally accorded the neighborhood ham radio operator. Today, in many social circles possession of a home computer is quite the status symbol, conferring an intellectual image on the proud owner. Suburban cocktail parties buzz with conversations about disk drives, VisiCalc, and modems. Home computing is finally "in." Those tireless experts who have been predicting every spring for the last twenty years that computers would soon be in every home, changing our lives forever, can at last retire from the field, vindicated. They were right, even if their timing was a little off. The public's demand for computers now seems insatiable.

VIDEO GAMES

To the intense surprise of almost everybody, during the 1970's the old-fashioned pinball parlor became the launching site of an entirely new industry which catapulted almost instantly into the multibillion-dollar category. Electronic arcade games, first introduced in 1972, generated United States revenue, mostly in quarters, of approximately $7 billion in 1982, according to estimates of the Electronics Industries Association. To put this amount into perspective, perhaps it helps to realize that the United States government produced the space shuttle for about $2 billion, that the exciting new industry of cable television made about $2 billion for the same year, or that the long-established motion picture industry's take was $2.7 billion. Once again, the fragile microprocessor has provided the opportunities and some daring entrepreneurs the will to establish an entirely new industry which never existed before and which has created thousands of jobs.

Initially the pinball industry welcomed micro-electronic technology as a design and manufacturing innovation. A few silicon chips conveniently replaced up to 70 pounds of motors, gears, wires, and levers on the old electromechanical machines, making them easier to produce, assemble, and distribute. Not especially concerned about potential competition for their own product, the pinball machine makers looked on with tolerance when Atari put out Pong in 1973. After a flurry of consumer excitement, the novelty seemed to wear off, and players returned to pinball. This industry thrives on novelty, however, and video games kept at least a foot in the door. Then, lo and behold! that foot let in an army in 1978, when the indefatigable Japanese struck yet again with Space Invaders, the most successful coin-operated game in history. No pinball company had ever made more than 30,000 units of the same game; today there are more than 300,000 Space Invaders machines in existence throughout the world—55,000 of them in the United States.

Space Invaders was the first arcade game to be programmed rather than elaborately wired and thus could be modified simply by changing the program. The famous Pac-Man soon followed, also from the Japanese. In 1980, Atari, not to be outdone, countered with Asteroids, which represented yet another technological breakthrough. In this game a player's space gun is not bound to earth but can fly, move every which way, free of gravity. And so it goes. Games like Qix, Red Alert, Venture, Omega Race, and hundreds of others are still packing them in. Although industry estimates describe players as overwhelmingly male (90 percent) and teenaged (80 percent), amusement centers, in keeping with their new title, now draw in a different crowd as well when secretaries, executives, and housewives line up to test their skills. In fact, the incredible popularity of the game with "children of all ages" intrigues social scientists, who are perplexed as to its significance. Some observers say that just as the apparently random activity of children at play always serves the larger purpose of training for adulthood, electronic game playing prepares the modern child for a world run by silicon chips. Mastery of electronic games may be the equivalent in computer literacy of learning the alphabet in print literacy.

If you are worried that the high-school-recess atmosphere of your local amusement center is unlikely to attract a particularly sophisticated crowd, there are other ways to develop computer literacy. The surprising success of the arcade games has provided the impetus for the development of other industries that never existed before and has changed the face of familiar ones.

The toy industry has long been with us, yet it has greatly changed in the past few years with the introduction of microchip technology. The toy industry was stunned in 1977 by Mattel's fabulous hit, Football, the first hand-held electronic game. A few Christmases ago anxious parents found the stupendously successful Merlin sold out across the continent. Between 1977 and 1979, the growth rate for all electronic games was 93 percent a year and 340 percent for the calculator-sized gadgets. But as consumers have become more sophisticated, that overall growth rate has shrunk to about 15 percent. Parents who have shelled out forty or fifty dollars for an electronic game only to find that their child puts it aside a few weeks later need only one such experience. Consumers now demand educational value and that elusive quality, playability, before putting down good money for what is, after all, only a game. During recent years a list of products offering proven playability and allowing ample scope to the imagination might have included, for example, Extex's hand-held version of Space Invaders, Milton Bradley's medieval fantasy game Dark Tower, and Mattel's Dungeons and Dragons.

In 1980 about 14 percent of all industry products used microchips. Indeed, the immense capabilities of the microprocessor have greatly expanded the educational sector of the toymakers' market, producing "toys" that are far more than mere games. In 1978, the semiconductor giant Texas Instruments leaped into this market with a new chip which could create a digital imitation of the human voice. Speak 'n Spell was the result, soon followed by Speak 'n Read and Speak 'n Math. Today small computers such as Mattel's Children's Discovery System and Entex's MAC teach spelling, arithmetic, musical composition, and even basic programming. Playskool has a computer, called Genie, for preschoolers which is as easy to use as a coloring book. There is no doubt that many parents, computer Neanderthals, are picking up plenty of computer basics as they learn alongside their children.

The popularity of electronic arcade games and the hand-held models sold to the home market is further reflected in the birth of two other interrelated products: video games and home computer video games. The distinction between the two is small but crucial. To play a video game, the consumer buys a module or console which hooks up to the television set and a joy stick with which the player guides movement on the screen. Individual game cartridges snap into the module. A home computer video game, by contrast, is fed directly into the computer by disk or cassette; the player manipulates the keyboard to guide movement on the screen.

Module home video games are now a $1.5 billion business. In

1981, 4.5 million module game machines were sold in the United States, a 125 percent increase over the previous year. About 10 percent of American households have a game console, and it is likely that the figure will increase. Atari, a division of Warner Communications, is by far the leader in this industry with about 75 percent of the United States market. In second place is the toy manufacturer Mattel, whose Intellivision line of video game products came out in 1980. Many of the individual games offered by the manufacturers are the same as those available in amusement centers or hand-held versions. Once the initial investment in a console has been made, however, it is obviously much cheaper for the consumer to buy video game cartridges, averaging about twenty dollars each, rather than to pump quarters into two-minute arcade games or to buy a new fifty- or sixty-dollar hand-held model for every new game. The arcade games offer higher graphic resolution and much greater audio fidelity. The module home video games offer greater economy, unlimited playing time, and fun for the whole family. No black leather jackets or punk hairdos need intrude, and the older folks can develop at their own, presumably slower, pace.

The far smaller, but potentially gigantic, home computer video game segment of the industry accounted for about $200 million in sales during 1981. In 1982 alone, however, more than 2.4 million home computers were sold and estimates have that figure doubling in 1983 to 5 million. It is not outlandish to predict, as many experts have, that every home with a telephone today, about eighty million households in the United States, will have a personal computer by 1995. It is in this sense that the home computer video game market has enormous potential. Atari cartridges are fun, but each offers only one game. For owners of home computers there are literally hundreds of programmed games, computer software, ranging in price from $3 to $130. The intelligence of the home computer permits much greater complexity in the games, which may range from strategies to simulations. In addition, the home computer allows the player to program his own games, a feature limited only by the programming skill and imagination of the operator.

The astonishing enthusiasm of the American public for electronic games in all their guises—arcade, hand-held, module home video, and home computer—reaches across age and class barriers. Partly, of course, the phenomenon reflects a response to novelty, but the popularity of the games is so sudden and so widespread that it must have deeper meaning. The public seems to be readying itself for the age of the computer we have all heard so much about. The schools are certainly not yet ready, nor is industry. In a sense the public appears to be finding its own way into the new world as best it can. Makers of home computers and video game consoles agree. In 1983 both Atari and Mattel started losing money on their video game divisions, as the public now seems to be turning directly to home computers. The line between game consoles and home computer sales is blurring. Atari, for example, spent millions developing the right software in the hopes of turning its video game computers into the computer of choice for wide home use. Commodore, maker of the PET home computer, has recently shifted its marketing emphasis back home after successful forays into the European market. Not long ago, Commodore President H. E. James Finke was heard to say: "Those games will move computer literacy into the home. A game computer can become a computer for educational programs and also for accessing data bases." So far the strategy seems to be working. Commodore's first half profits in 1983 more than doubled its previous year's. Commodore also led the way in a savage price war which saw the cost of its cheapest home computer drop from $199 in December 1982 to $89 in June 1983. After suffering stunning losses, Atari too finally cut its prices to $99. Coleco, a video games success, now offers an entire home computer system, including a letter-quality printer and a high-speed tape memory for a mere $600; others charge four to six times this price. The future will see even more marketing strategies develop in this industry, for there is no doubt that America's love affair with the computer has just begun.

THE GAMES INDUSTRY: JOBS

The video game may well be a mere flash in the pan, a novelty soon to be discarded. The utility of the game in providing many with the rudiments of computer literacy, however, will last. In fact, most probably video games themselves will continue to exist in some form or other if only because they can be such a valuable tool in teaching the young.

Although it seems unlikely that the video games industry will maintain its recent phenomenal growth rate in the years to come, it is too soon to say there is no future in this area. The sixteen-billion-dollar-a-year American consumer electronics industry offers a host of new technologies, including, among others, videocassette recorders, video disk players, and digital recordings. At the moment the electronic games are the biggest dollar grossers. But the tiny microchip, which has already made millionaires out of hundreds of clever entrepreneurs, has potential applications undreamed of today. The fragile microchip, each year smaller, more powerful, and cheaper, literally is limited only by human imagination in its application. Perhaps even now some visionary has cooked up a new use for the chip which, like the video games, will spawn a multibillion-dollar industry almost overnight. Since that visionary has not yet appeared, however, we must content ourselves with a look at the amazing jobs and incredible opportunities that already exist in the video games industry.

Arcade video games are part of the industry started by pinball over fifty years ago, whose members are organized as the Amusement and Music Operators of America. Every year the organization holds a convention at which at least 150 new pinball and video machines are unveiled to the public for the first time. These machines are created by the undisputed stars of the video game world, the designers. For the most part, they are young, male, and fanatical in their devotion to game playing. They also do very well financially, pulling down $50,000-to-$100,000-a-year salaries. Some free-lancers may earn even more if they are exceptionally good and work very hard. Some are engineers, some are college dropouts, some are high-school students. All share a single-minded devotion to games. While engineers concentrate on creating sharper lines, more elaborate methods of annihilation, or aliens with brighter colors, the designers produce the overall world of the game and its ultimate purpose. The designer must make the game feel right. The designer looks for "clank," as it is called in pinball, that something which makes a game awkward or uninteresting. It is an indefinable quality which can bore the player and ruin a game that cost $250,000 to develop. If you have to ask what clank is, you'll never know; if you instinctively recognize it, you may well have what it takes to be a designer.

Game designers of the future will undoubtedly come from a suprising variety of areas. Recently, as interest in arcade games has fallen off, the fate of the industry has hinged upon the creativity and imagination of its designers. Dragon's Lair, a new kind of game, is once again packing in players who have become bored with decimating electronic blips. An innovative use of an established technology, Dragon's Lair employs a video disk to project an animated color image of almost movie-like quality onto the screen. Introduced in mid-1983, its success was instantaneous, providing new hope to a beleaguered industry. Dragon's Lair also shows that cartoon animators as well as special effects artists or technicians have a place as games designers. Designers of the future will have an even more diversified repertory of skills.

Working alongside the designers are the programmers. They, too, are usually games buffs who positively enjoy the tedious two-month process of telling the machine how to play the game. It is the programmers who must fit everything into two minutes and into the limited memory space chosen. A machine with four chips would contain 32,000 bits of information. Adding any more would raise the cost of each machine by fifteen dollars, an expense which could add up to hundreds of thousands of dollars in the course of production.

Then there is the business side of the industry. Each machine is sold at a cost of about $4,000 to a distributor. The distributor is responsible for finding the outlets—supermarkets, bars, amusement centers, or any other legal spaces. Many cities have municipal restrictions on where the machines can be placed, when they can be played, and sometimes on the age of the players. Usually the site owner leases the machine from the distributor, who takes it back for reprogramming after about six months, when the boredom factor has set in. The business is extremely lucrative. In 1981, for example, Atari's revenues on sales to distributors doubled the previous year's to $415 million. As for the individual machine itself, a hot item in a good location can easily collect at least $500 a week. Industry openings are listed in the United States industry's trade journal, *Play Meter*, and in *Canadian Coin Box* for those north of the border.

There are spectacular success stories in the games business. In *Running Wild* Adam Osborne tells the story of Nolan Bushnell, the young electrical engineer who designed video Pong and started a whole new industry. With neither the financial resources nor the management background others might have thought necessary, Bushnell, a product of the fast-paced world of micro-electronics, decided to found his own company. He called the company Atari. Within five years an entire industry had grown up around him, and in 1976 he sold his company to Warner Communications, becoming, in his early thirties, a multimillionaire. His story, as Osborne points out, is hardly uncommon in the micro-electronics industry. There nobody knows or cares about job security. There risk taking is the way to do business, and the chances are good for incredible success. Established companies, in other industries, mired in bureaucratic procedures, adapt slowly to changing markets and innovative products. In the highly competitive micro-electronics industry, a technological breakthrough or a new application for the microchip may mean an advance of only a few months or even weeks over the competition. So the players must learn to act quickly, reap the profits, and retreat—fast. It is normal procedure for the smaller entrepreneur, like Nolan Bushnell, to be bought out by a larger company once his business is a going concern. This road to instant wealth is a well-trodden path in the home computer game segment of the micro-electronics industry as well.

The designers are also the superstars of the home computer game industry. Unlike his opposite number in the arcade industry, however, the average home game designer dreams up the games and programs them too. Since Atari and Mattel are the only major companies in this nascent industry, many young game creators work free-lance, incorporate themselves, or found their own companies. Success stories abound. James Nitchals and Barry Printz, for instance, both twenty years old, are the owners of Cavalier Computer Corporation of Del Mar, California, which had about $240,000 in sales in 1982. While still in high school, Nitchals wrote Asteroid Fields, the company's first product, on his Apple computer. Then there is Roberta Williams' game, Mystery House, which she and her programmer husband marketed initially through advertisements in game magazines. When the first month brought in sales of $11,000, they founded On-Line Systems, with 1982 sales hovering at the $10 million mark. David Crane, while still in his twenties, left Atari in 1979 with three other game designers to found Activision, with 1982 sales of $65 million.

Royalties on a game can run from 25 to 35 percent. A popular game can sell 25,000 copies, and some sell as many as 50,000. The designer, then, may make between $75,000 and $175,000 for a game taking from four to eight months to program and package. Some designers work for a company on a salary basis with the proviso that they get a cut of the profits. Twenty-year-old John Harris, creator of Mousattack and Jawbreaker, made $300,000 that way in 1982.

Like their arcade counterparts, most successful designers are game enthusiasts first and then competent programmers. Many are too young to have finished their studies, others are self-taught. The first qualification for entering this field is a love of games; the second is a desire to make money and a willingness to take risks. The rest looks easy—but of course it's not.

As the arcade market begins to seem saturated, the home computer game market, as yet almost untapped, must appear more attractive to large companies all the time. In addition, home computers offer possibilities for marvelously sophisticated games. Adventure, for example, long a favorite of computerites whiling away the long night hours on huge lab and university computers, is in the library of both Atari and Microsoft. Flashing words rather than pictures onto the screen, Adventure describes an imaginary situation in graceful prose, leading the player down sparkling streams and through the forest primeval to either treasure or a dragon. Not incidentally the enchanted player learns deductive logic while struggling to capture riches. Indeed, the educational possibilities of home computer games are almost endless.

At one time Massachusetts Institute of Technology scientists and educators were working on video games that taught children the laws of motion and logic. While no commercial company has yet packaged such a game, the time when one will do so cannot be far off. Computers proliferate throughout our society, moving inexorably into the home, and schools must prepare children for the future that is to be theirs.

CHAPTER FOUR

DOMESTICATING THE COMPUTER

During its brief lifetime, the microchip has broken all the marketing rules dictated by the experience of previous industries and plain old common sense. Usually introducing a new technology into the home poses a classic dilemma. Until great numbers of consumers find the product useful, prices remain high and quality low. While prices remain high and quality low, few people want to buy the product. In its first few years of existence, television faced this difficulty. Not many people had sets, so viewing time was limited to a few hours a week. All over America status-conscious television owners watched test patterns on their tiny black-and-white screens while the rest of the country bided its time. Eventually, of course, the market evolved, and quickly thereafter screens grew in size, and then color made its appearance. The telephone had the same problem, as did the automobile. It makes sense that this should be so; it seems reasonable.

What seems reasonable or makes sense doesn't necessarily hold true in the micro-electronics industry, built into a gigantic market force in just a few years by people so ignorant of common business practice they didn't realize it couldn't be done—until after they had already done it. They may have known next to nothing about marketing schemes, the importance of packaging concepts or advertising schedules, but they did know about something the average businessman didn't: the incredible chip and how to make it work. Today we have all learned how a single three-dollar chip or two, joined together, can rival the computing capacity of the early multimillion-dollar monsters. In the early days of the microprocessor, home hobbyists were among the first to discover its potential and, as the ads appeared in obscure computing journals, stampeded to buy kits, boards, or whatever they could get their hands on. Shortly the wider world found out, too, and the microcomputer revolution began.

Although all this sounds like a tale of now impossible feats from the early days of the Industrial Revolution, of robber barons making enormous fortunes, it all started in 1975.

In that year Micro Instrumentation and Telemetry Systems (MITS) of Albuquerque, New Mexico, brought out the first true microcomputer, designed around Intel's still-popular 8080A microprocessor, and marketed it in kit form as the Altair. Unprepared for the deluge of orders, its makers sold 2,000 kits that year by pushing their manufacturing capacity to the limit. Shortly afterward, another company, known as IMSAI, marketed a similar product; in fact, their kit was purposely designed to be compatible with the MITS kit. Now outside designers of peripherals and writers of the all-important software no longer had to create for each separate company's product and their efforts could become economically feasible.

Less than two years later, the industry, for that was what it had already become, stopped producing kits. Turning away from its hobbyist beginnings, it quickly reoriented itself toward the far more lucrative business market. (See Chapter 7, p. 151.) Back in 1977, the home market seemed to be a fizzle. Although small enough and cheap enough to interest the home buyer, the microcomputer seemed simply too powerful a tool for a public that quaked at the very word computer. Who needed a computer to balance a checkbook or keep the Christmas card list? Who had time to learn how to program the silly thing? For the home market the microcomputer looked like a bad case of technological overkill.

Then along came video games. With the extraordinary interest of the American public in these games, the mass market, almost written off by microcomputer manufacturers, began to beckon promisingly again. A few years can be an eternity in this speedy industry: this time around the microcomputer industry had changed dramatically. The microcomputer field is now packed with competing companies, most with experience in meeting the needs of both small and large businesses. Today's microcomputer companies originated in every possible industry: from semiconductors to integrated electronics, mainframe computers, scientific instruments, home computers, and, of course, video games. These firms have already developed hardware specifically for the non-technical user. The home computer user now benefits from high-level languages that have evolved during the past few years. And, most important, software for the home user now fills libraries.

What exactly is this wonderful machine? How does it work? What is all that mysterious jargon about?

WHAT IS A MICROCOMPUTER?

Defining a microcomputer adequately is a challenging task. Essentially a microcomputer is a computer, a general purpose problem-solving machine, which is built around a microprocessor. Even that limited definition, however, raises difficulties. Modern computing machines fall into roughly three categories: mainframes, minicomputers, and microcomputers. For our purposes we can disregard the mainframe since that term refers to the large, complex machines made by giants of the industry like IBM and DEC and which cost hundreds of thousands of dollars. These are the machines, descendants of the original ENIAC, which inspire such fear and trembling in the general public. From countless movies we know them as rooms or walls full of blinking lights, emitting strange, extraterrestrial noises and tended by a white-coated priesthood of data-processing professionals in an absolutely sterile environment. Such machines are most likely to be found in very large institutions: banks, government offices, and multinational corporations.

The established mainframe manufacturers had, of course, nothing whatsoever to do with the development of the semiconductor industry's microprocessor. These companies, however, did scale down the larger machines to produce the minicomputer. Minicomputers, powerful machines usually found in business, are made by IBM, DEC, Hewlett-Packard, and others. Today's minicomputer also makes use of microprocessor circuitry. In fact, minicomputers and the better microcomputers are now about the same size, look alike, and can perform more or less the same functions. Thus the distinction between them becomes blurry and finally rests on price: the smaller minis are in the price range of the more expensive micros; the mini usually starts at about $16,000, far out of the range of the home buyer. A microcomputer for business use may cost from $4,000 to $20,000, depending on the value of extra equipment or software needed. A microcomputer for home use could cost anywhere from $50 to $4,000, again depending on the value of the extras.

A stripped-down microcomputer is a fairly straightforward machine. Its constituents fall into two broad categories, hardware and software. Hardware means the actual physical machine, and software means the programs, or instructions, which allow the machine to do something. In common parlance people usually mean the hardware when they refer to a microcomputer or indeed to any computer. To give an inexact comparison, if a computer were a television, we might say that the television set was the hardware, while *The Evening News* or *Sesame Street* was the software. The reality of a computer is slightly more complicated, as we shall see.

In terms of hardware, all computers, even the giant mainframes, typically consist of three basic parts: an input device, a central processing unit, and an output device. Technically the input and output devices are really peripherals, so called because they are peripheral to the functioning of the central hub of the machine, the central processing unit. There are also other types of peripherals. In practice, these parts are often physically separable and can be changed, upgraded and downgraded, just like a modular stereo system. To further confuse matters, some people mean only the central processing unit (CPU) when they say "the computer," while others may mean only the input device or only the output device. Whether we like it or not, computer jargon has entered our language, and it often leads to confusion outside the specialized world where it was developed. Schematically the computer looks something like the figure on page 44. The input device feeds the data and instructions to the CPU. In the old days on the big machines, the input device punched cards and fed them into the machine's innards. Today, the most popular input device is usually similar to a normal typewriter keyboard. There are a number of keyboard variations, either simpler or more complicated than the familiar QWERTY. In most, but not all, cases a full-fledged home computer has a typewriter-like keyboard.

The output device displays, prints, or records results of the computations performed by the CPU. In most home computers, the output device is a video display screen, sometimes called a CRT, short for cathode-ray tube; the CRT may well be the home's own

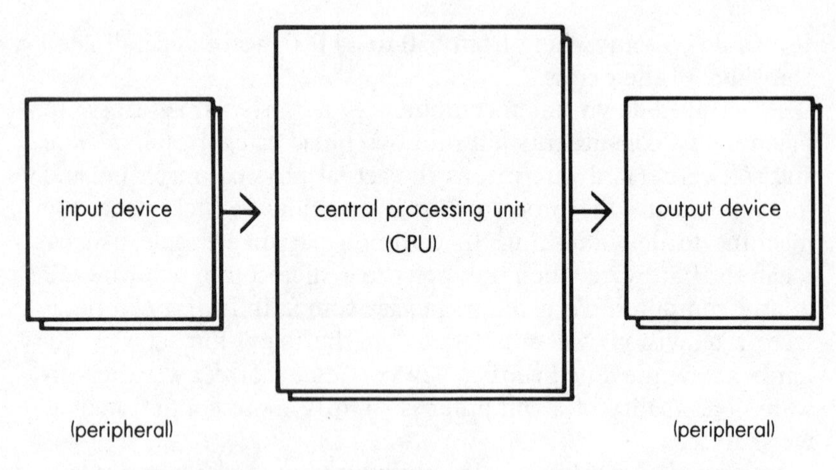

FIGURE 4.1

television screen. Since the home television set is usually in demand for television viewing, a specialized CRT is often part of a home computer purchase. Many home computer CRTs have black and white capacity only, but some, even the most economical models, have extensive color capacity and superb graphics. The legacy of the video games, which rely on vivid colors and graphics for exciting effects, is most evident here.

The CPU contains the "brain" of the machine, which does all the real computing. In the home computer the heart of the CPU is the microprocessor, a microchip which is itself sometimes called the microcomputer. The CPU may have more than one microchip. As we have seen, a microprocessor consists of three parts: a control unit, which executes the instructions in the correct order; an arithmetic-logic unit, which actually performs the instructions; and a temporary memory storage unit, which allows the computer to remember what it is doing at the moment. The CPU may also contain other chips, called memory chips. These chips perform several functions, depending upon the power and capabilities of the machine. Some memory chips are loaded on in the factory with some permanent internal programming. These instructions telling the machine how to compute are the read-only memory (ROM). In effect, they are software that is internal and permanent; they are called "firmware." Other memory chips store data and can "write" as well as "read." These give the computer more specific information for reference and comprise the random access memory (RAM). Other data may be held externally. The size of the processor memory helps to determine how fast the computer can handle its tasks. Memory size is measured in bytes; a byte is the amount of memory needed to store a single alphabet letter or a numeral. For a home computer, a processor memory of less than four thousand bytes (4K) would be considered small. Processor memories of sixty-four thousand bytes (64K) and more are available. Now we can see why the earlier comparison of a computer to a television set is inexact. In a computer some of the programming instructions are located within the hardware itself. In a way, that would be like having Big Bird and Cookie Monster actually existing inside your television set, while Kermit and Oscar the Grouch had to be broadcast.

Another indispensable part of the home computer is the interface. An interface is an electronic circuit board which makes an input or an output device compatible with the CPU. The interface(s) may be part of the CPU or part of the peripheral device. Frequently some of the internal programming memory of the computer, contained on a chip, is part of the interface. Schematically a more detailed picture of a home computer is shown in the diagram on page 46.

PERIPHERALS

Technically any part of a microcomputer aside from the CPU is a peripheral. In practice, however, many micros have an input device, typically a keyboard, and an output device, typically a CRT screen, housed together with the CPU. This makes it hard to think of them as separate devices. Without these input/output devices, a computer is blind and mute, unable to take on problems or communicate the answers to the outside world. So peripherals are extremely important items which add to a microcomputer's flexibility and range of applications. Peripherals also add to the cost of a microcomputer.

The simplest input device is a keyboard; it looks exactly like a typewriter keyboard but has no types and, of course, no paper.

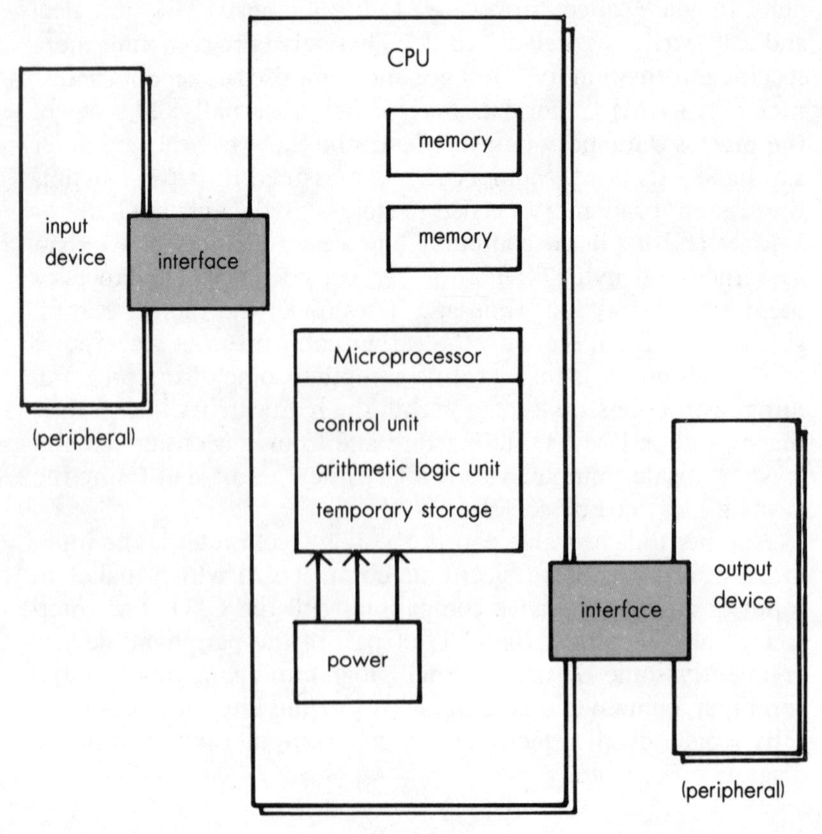

FIGURE 4.2

The CRT displays what is typed in. Keyboards come in three general styles: the full stroke keyboard is the full-size, big-computer type model; the calculator keyboard has the feel and limited stroke of keys on a pocket calculator; the diaphragm keyboard, the least expensive model, is a thin sheet of metal with raised bubbles for keys.

In order to cut costs to the bone, many manufacturers offer CPUs which use the home television set as the output device. This trick can be accomplished by building a radio-frequency (RF) modulator into the CPU. All computers with an RF modulator have a television/computer selector switch which prohibits the television set from performing both duties at once. However, there are distinct disadvantages to using the home television set as the output device. A computer CRT has a finer, more expensive tube than the television set, providing higher resolution and greater clarity of detail. A CRT display accepts entire "pages" of data in a fraction of a second and holds them in view for as long as necessary. With adequate circuitry and programming, a CRT display can scroll the data upward and downward on the screen so that the user can locate something quickly or go back and read something again.

Many CRT units incorporate a keyboard; this combination is called a terminal. Any combined input/output device primarily for communicating with a computer is known as a terminal. Other types of terminals do exist, such as teleprinters or even modified electric typewriters, but these terminals are usually found in business. Some home computer keyboards, unlike business-oriented machines, have the ability to generate complex musical sounds, ranging from a single short beep to multipart tones in variable frequencies far outside the range audible to the human ear.

Other peripheral gadgets function as input devices for a home computer, including joy sticks, game paddles, and pistol grips, all of which are used for games but which can also serve as inputs for graphics entry or musical composition as well. A light pen or electronic pointer can make changes directly onto the CRT screen; the computer understands which part of the display it indicates.

Printers are popular output devices. Due to the cost, a printer is usually added to machines designed for business use, but it can be an invaluable peripheral for the home computer because it provides printout listings to facilitate the inevitable task of program debugging. The cheapest printers use an electrostatic printing process and require a special chemically treated paper. Slightly more expensive is the dot-matrix printer, which operates at very high speeds. Another, even more expensive, alternative is the daisy-wheel printer, which produces typewriter-quality print—usually far finer than the average home-computer user needs. Printers come with numerous options, such as upper and lower case characters, with the cheapest only in lower case; line widths which can vary from 20 to 132 characters with 80 as the average; and various speeds.

Home-computer manufacturers also supply a variety of control units. One such unit can cause signals from the computer to switch electrical power in the home off and on, controlling lights and appliances. Another can act as a burglar alarm; yet another is a smoke detector or a medic alert. All these options add to the price of the home computer.

Another peripheral for a computer is a modem, short for modulator/demodulator. The modem enables a computer to connect to an ordinary telephone line and thus to various commercial data bases, banks of information stored electronically. The modem converts the computer's output data pulses into audio tone signals understandable to a digital telephone, such as a Touch-Tone, and then converts the incoming audio signals into data pulses. Usually the handset of the telephone fits into a pair of rounded openings at the top of a small modem cabinet; this type of modem utilizes ordinary telephone (voice) lines for occasional transfer of data (information in the digital mode understood by computers). Sometimes in the more sophisticated business micros a modem permanently hooks into a data line, which is used exclusively to transfer digital information; a data line is far faster than the ordinary voice line.

A most important microcomputer peripheral is a mass storage system, commonly an audio cassette recorder or sometimes the more expensive floppy disk or diskette systems. Aside from acting as a file for data that would overload the computer's internal memory, this external computer memory is also computer-compatible, allowing information to move easily to and from the machine without tedious keyboarding. The cheaper models are cassette drives which accept ordinary audio cassettes. Since the CPU processes data much faster than the input drive can feed it, the most practical method is a dual cassette drive. One cassette inputs the data, while the other outputs the results as the final bit of information enters the computer.

Users with more sophisticated filing needs, especially in business, generally opt for disk files, using floppy disks, or diskettes. Disk systems cost more than cassette systems, but, unlike tape, floppies cannot break. Disk drives are available in dual or double dual models. (Floppies are discussed in detail in Chapter 7, see p. 171.)

The cost of a home computer is extremely variable and depends to a large extent on the choice of peripherals. A sophisticated microcomputer system currently costs less than a small car. In terms of cost and computing capability, there is no doubt that a microcomputer is a truly democratic device, well within the reach of the majority of the population. Anyone who can afford to buy and run a car can afford to buy and run a microcomputer. Although the price is falling every year, the microcomputer is probably not yet in the "must have" category for the average person. As of late 1983, some of the leading microcomputer products offered included the following:

Radio Shack TRS–80 Model IV provides a basic CPU with a 16K memory, a twelve-inch video display, and a keyboard for $999. A tape cassette unit adds $59.95 to the price, and a floppy disk unit costs $249. To add a daisy wheel costs between $799 and $1,995. Modems range from $99.95 to $149.95, and an appliance control unit runs $69.95.

The Timex Sinclair 1000 basic unit is an extremely compact CPU with a diaphragm keyboard and a 1K memory for under $50. A separate module at $99.95 expands the memory to 16K. The unit uses the home television set as an output device.

Apple's starter system, the Apple II E, includes a CPU with 64K memory; a keyboard; a video display; a disk drive; and a copy of Logo, the company's programming language, all for under $1,500. Part of the immense success of the Apple line of microcomputers stems from the wide range of software available for the system.

SOFTWARE: USES OF THE MICROCOMPUTER

The second major constituent of a home computer, aside from the hardware, or the machine itself, is the software—the instructions, or programs, which tell the machine what to do. Any computer is useless without software. Programs imbue the computer with intelligence; they can make the machine do almost anything you want it to do. With appropriate software, a microcomputer can prepare your tax form, file your favorite recipes, make mailing lists, correct your terrible spelling, teach the kids math, help you decide which new car you want or whether you really need one at all, act as a home burglar alarm, play poker with you, turn on the coffeepot in the morning, remember everybody's birthday, give you the stock market reports. In fact, software is so versatile that this list could be expanded almost indefinitely.

Anyone with the easily acquired requisite skill can program a home computer. In the early days when microcomputers were sold in kit form, the home hobbyist buyers themselves frequently programmed the machines for specific applications. For this type of knowledgeable buyer, programming the computer is part of the fun of owning it. Today, however, the average home-computer buyer knows little and cares less about programming and wants a machine that does something. For this kind of purchaser, software is prepackaged. When a home computer appears with all the peripherals you want and all the software you need for your particular purposes, it is called a turnkey system. That means that it is like buying a car with a full tank of gas: you just turn the key, and off you go. Getting a home computer to that point, however, is a little more complicated than driving away in a new car.

Essentially a commercial software package consists of programs for the computer and documentation for the user. Documentation usually comes in manual or book form and explains how to use the program and what machines it was designed for and gives other necessary information. Programs, which come packaged in magnetic form, either tape or disk, or in cartridge form, are of four types: operating, languages, utility, and applications. Operating systems, internal in most home computers, direct the program's commands to the appropriate hardware. Language programs provide a means to put the user and the machine in communication. Utility programs are general purpose instructions or a set of instructions, directing the computer to perform a particular task which is frequently repeated, such as sorting a list. Applications programs direct the micro to do something specific, such as alphabetize a list of names or do a payroll. Here we will concentrate on applications programs and languages.

Most packaged applications programs are in the interactive mode. This means that the program is written in such a way that it prompts the user with clear requests for information or actions. As an example, suppose you have a program that makes a simple list of the names and addresses of your friends and business acquaintances. Now someone has moved, and you want to change the address. With the program in place, the display screen will print something like this:

```
WHAT DO YOU WANT TO DO?
1. ADD NAMES/ADDRESSES
2. DELETE NAMES/ADDRESSES
3. CHANGE NAMES/ADDRESSES
4. RUN LIST
5. END PROGRAM
ENTER SELECTION:
```

Since you want to change an address, you depress number 3 on your keyboard. Now number 3 will appear on the screen next to ENTER SELECTION. You then hit RETURN or its equivalent on your keyboard to continue. The screen now prints:

```
WHAT INFORMATION DO YOU WANT TO CHANGE?
1. NAME
2. ADDRESS
3. TELEPHONE NUMBER
4. END CHANGES
ENTER SELECTION:
```

This time you enter number 2 for an address change, and the screen prints:

```
ENTER NEW STREET ADDRESS:
ENTER NEW CITY:
ENTER NEW STATE/PROVINCE:
ENTER NEW ZIP/POSTAL CODE:
LAST ENTRY (Y OR N)
```

After entering the details, you tell the program this was the final entry by typing in Y for yes. The program then displays the previous WHAT INFORMATION DO YOU WANT TO CHANGE? page. You type in number 4, END CHANGES, and the machine returns to the first page, WHAT DO YOU WANT TO DO? You enter number 5, END PROGRAM.

Obviously this procedure is not very difficult. Reading about how to make the change takes much longer than actually making it on a computer. The procedure can be so simple because the program uses a high-level language. Although the term sounds forbidding, a high-level language is merely one closer to English or any other human language. A low-level language is one closer to the machine language of *0*'s and *1*'s. To produce the type of display shown above, the program employs a high-level language, which another program, already in the computer, translates downward into machine language.

Originally users communicated with the computer in low-level languages. Programs written directly in machine language can eliminate the need for translating and so save computer memory space and running time. The great disadvantage of machine languages, of course, is that only those skilled in the languages can communicate with the machine. In the beginning of the computer era, when only computer experts touched the machines, language wasn't too much of an issue; but later, as nonprofessional users in increasing numbers clamored to use computers, specialized languages were developed. A physicist or an economist, for example, does not want to waste precious time translating everything into machine language. High-level languages save human time by making the machine do the translating and make it possible for the ordinary person to use the machine without the intermediary of computer professionals. Some high-level languages are ALGOL, FORTRAN, COBOL, PL/1, RPG, and PASCAL. Each title is an acronym, generally indicating the application of the language. Thus ALGOL stands for Algorithmic Language and FORTRAN for Formula Translation; both have scientific and engineering applications; COBOL means Common Business Oriented Language and is used in business. BASIC, or Beginner's All-Purpose Symbolic Instruction Code, originally designed to teach programming concepts on the big mainframes, is the most common program language for microcomputers.

A computer language differs in significant respects from a human, or natural, language. Most computer languages consist of a small number of command verbs, like *print, run,* or *go to,* and a few comparison or conditional statements, like *if . . . then*. Since the vocabulary is limited, the language is easy to learn, yet since the vocabulary is also very general, it can be applied to a variety of highly diversified activities. Unlike natural languages, where words frequently, even usually, have multiple meanings and permit of several interpretations, computer languages are universal: every item in the vocabulary has just one meaning, so that everyone understands and interprets it in the same way. Separate computer languages were developed in order to enter information for a specific application by the most efficient method. Different purposes require different languages, yet each language can be thought of as a subset of a universal computer language. No parallel in natural languages exists. The idea, for example, that Chinese might be more efficient for communicating business information, while English is preferable for a discussion of religion, is absurd. The specialized vocabulary of a computer language allows the programmer to express his logic more efficiently. Thus while you don't need to know COBOL or some other business-oriented language in order to program a computer for business applications, knowing the language would save you a lot of trouble.

Although computer languages are very different from natural languages, a high-level computer language displays many similarities to a natural language. BASIC, for example, looks much like a kind of simplified English; it is readily understandable. Suppose you were taking a trip and you wanted to estimate how much it would cost you to go by car. An entire program in BASIC requesting that information would look like this:

```
100  PRINT "HOW FAR WILL YOU GO? (M) ="
110  INPUT M
120  PRINT "AVERAGE GAS CONSUMPTION (G) ="
130  INPUT G
140  PRINT "COSTGALLON ($) ="
150  INPUT C
160  X = (M/G)*C
170  PRINT "IT WILL COST YOU ($) ="
180  END
```

With a moment's thought any English speaker can decipher the meaning of this program. A program proceeds in a commonsense way. To find out the cost of the trip, you need to know the distance; the car's estimated gas consumption at, say, fifty-five miles per hour; and the cost per gallon of gas. Simple deduction will tell you that the asterisk in the formula X = (M/G)*C replaces the conventional x sign (because the computer might confuse it with an *X*).

If this program were actually run on a computer, only the information appearing after PRINT and INPUT would come onto the display screen. The formula itself—the method of calculating the result—would not appear on the screen. The computer would print only the answer. As can be seen from this extremely simple example, it is the program which tells the machine what information to ask for and in what order; most important, it is the program which tells the machine how to calculate the answer. All this information, ready for instant use, is contained in a commercial software package.

THE HOME COMPUTER INDUSTRY: JOBS

Market forecasters predict phenomenal growth rates for the home-computer and associated industries during the 1980's and 1990's. The potential domestic market is the 80 million American households which now have television. Some forecasters go so far as to say that 80 percent of these households will have a home computer of some sort by 1990. Considering that only about 2.4 million home-computer units were sold in 1982, the prediction of an increase to 80 percent of the market, or 64 million units, does, certainly, augur well for the microcomputer industry. That figure, of course, does not include the export market or the business market, which today is much larger than the home market. According to SRI International, home buyers should outnumber small business buyers by 1985. The personal computer industry has been growing at a healthy 40 percent annually for the past several years; sales topped a billion dollars in 1981. According to *Infoworld*, a weekly newspaper about microcomputing, the industry's global total is about three billion dollars when sales of software, support systems, and peripherals are included. In 1982, according to Dataquest, a California research firm, more than 100 companies sold 2.8 million units for 4.9 billion dollars. Analysts expect this figure to swell to fourteen billion by 1985. Numbers like that cannot be ignored. If the predictions are on target, this industry will indeed be "running wild" throughout the 1980's and 1990's. This would generate thousands of new jobs in manufacturing, in the support or service area, and in spin-off industries; it would also open entrepreneurial possibilities by the hundreds. The question is, though, are these figures based on reality or not?

Though the home market has, as yet, hardly materialized, there are sound reasons for the forecasters' optimism about the future of the industry. For one thing, the unexpected success of video games with widely divergent groups shows that the general public is manifestly fascinated by computers. Although computerphobia is still epidemic on this continent, especially among the middle-aged and older, the spread of video games points to an increase in computer literacy. So, too, does the burgeoning number of microcomputers in business, both large and small; after all, ordinary working people of all ages are learning to operate these machines and integrate them into their business lives. Once computer literacy becomes widespread, it is axiomatic that the sale of home computers will rise.

Another point to consider is that in this world of spiraling price increases, home computers are and will continue to be one of the few genuine bargains left; prices are constantly falling, while quality is improving with lightning speed. The modern micro has computer capability equal to the largest machines of the early 1960's. Those machines cost millions; today's equivalent home computer costs about as much as a quality stereo system. During the last five years, the number of components on a chip has increased by a factor of 100; it is likely that micro-electronic complexity will increase 10,000-fold by 1990. Some industry enthusiasts even speculate that the capacity of the world's largest computer, Cray I, which cost nine million dollars, will be available in microcomputers toward the end of the decade. Since developments in the micro-electronics industry have been so miraculous, even experts are reluctant to negate any speculations, however fantastic, about its future. No matter how spectacular the industry's advances, though, a public ignorant of computers cannot appreciate them. But as knowledge about computers increases, more and more people will become aware of the industry's achievements. In time, a greater percentage of the population will understand how the machines work and their potential applications around the home. Even now, people are coming to realize that the microcomputers' bargain-basement prices represent a real steal for the consumer. As a significant percentage of the public gains computer literacy, this industry will expand dramatically. It could easily happen during the 1980's. By the 1990's it will be all but inevitable.

MANUFACTURING SECTOR

Already the personal computer industry, small though it is, has generated many jobs so new that even ten short years ago they simply did not exist. Throughout the eighties, into the nineties, and beyond, the types of employment created will snowball as the industry matures. Although we cannot know what specific jobs will exist ten years from now, we can identify the sectors in which they will appear. At the moment a *manufacturing sector*, which has two separate and distinct branches, is rapidly expanding. One part of this sector produces hardware of all types; the other produces software. The two branches of the sector, though often represented by completely different companies, are totally interdependent; one could not exist without the other. Another indispensable growth area is the *support sector*, including service and repair, managerial and marketing groupings. Yet a third division is what might be called the *resource sector*. Potentially the largest of all the sectors, it has both an educational and an informational aspect and will generate a multitude of salaried positions and entrepreneurial possibilities. This sector will experience its greatest development as the industry matures in the 1990's. It is important to realize that when microcomputers become fixtures in even 30 to 50 percent of the private homes in the United States, the number of people employed as a direct or indirect consequence will be truly astounding.

HARDWARE. Since a microcomputer has so many applications, it is not surprising to find that the hardware manufacturers are dividing themselves up according to their target markets. Thus some companies gear themselves to the high end of the market, producing the more expensive and sophisticated machines for use in business; other companies concentrate on the still emerging low end, marketing the far cheaper personal computers. At the moment the situation is still fuzzy, but the battle lines are being drawn. Until recently, most micro manufacturers were smallish outfits like Apple, Atari, and Commodore which originated with the industry. The exception was the giant Tandy Corporation, a retail electronics firm, producing the popular Radio Shack TRS line of machines. Today, however, the boom in micro sales, especially in business, has drawn larger, more diversified corporations, like IBM, Wang, Xerox, and many others, into the fray. The immediate result is vastly increased research and development, greater production, more specialization, and extremely keen competition. The end result could well be the demise of the smaller companies.

Industry analysts expect that hardware sales of microcomputers, defined as those selling for less than $10,000 and designed for use by one person, will climb from $1 billion in 1981 to around $9 billion in 1985. Total 1981 sales were about 1.2 million units, representing a 400 percent increase over the 278,000 units sold in 1978. The Dallas-based research firm Future Computing confidently expected 1983 sales to total 5 million units, representing retail sales of $2 billion. The almost untouched European market also promises much. By the end of 1982, there were only 1.7 million personal computer units in European homes and offices, but the Norwalk, Connecticut, market research firm International Resource Development predicts that figure will be 18 million by the end of the decade with West Germany the largest market. IBM, Radio Shack, and Apple are among the many American hardware manufacturers successfully competing in this as yet untapped market.

Apple, Radio Shack, and Commodore, the established companies in the industry, were the original developers of the home and business markets. Texas Instruments, the leader in semiconductors, and Hewlett-Packard, the largest maker of scientific instruments, are selling mainly to the scientific community and to what they call analytical professionals, such as financial analysts, planners, and engineers. Digital Equipment and Data General, primarily minicomputer manufacturers, target business users. Nippon Electric's micro line, which captured 7 percent of the 1981 market, reminds everyone of the ever-present Japanese threat. The micro market is now so attractive that, in a most interesting development, even mighty IBM, the most powerful world force in mainframes, dropped its haughty disdain for the small machines and put out a personal computer called simply PC. Although introduced only in August 1981, the PC nevertheless garnered 4 percent of the entire market for that year. By mid-1983 IBM had captured 21 percent of the market, a staggering feat in so short a time. IBM also planned to market a $700 personal computer, first code-named Peanut, now called PCJR, in late 1983 which will, according to company spokesmen, offer the best performance on the market for the price. Other micro manufacturers must take that to heart: IBM's famous name and recognized excellence in the computer field mean very tough competition ahead.

Greater competition, more specialization, and increased production are, of course, excellent news for job seekers. For the near future, prospects in hardware manufacture are very good. Some of the technical jobs available have already been touched upon in the section on the semiconductor industry. Apart from the jobs involved in the fabrication of microprocessors and memory chips, electronics engineers, electrical engineers, and specialists in testing and instrumentation will be in demand. As the years go by, the industry will require engineers educated specifically in the techniques of microcomputer production. However, there is some doubt whether the supply of adequately trained engineers—not to mention highly specialized engineers—will be sufficient in the 1980's and 1990's. If not, there will be a greater demand for teachers of these disciplines at the university level. If the supply of engineers is inadequate, then almost certainly the better-prepared technicians will be upgraded. This, in turn, will result in an increased demand for teachers at the junior-college and technical-school level.

Production for the masses, combined with increased specialization, leads to more intense research and development (R & D) efforts on the part of manufacturers. The latest figures indicate that industry R & D costs are rising significantly. In 1977 Tandy spent an estimated $150,000 developing its first personal computer; its latest TRS-80 Model 16 cost millions of dollars. Part of these rising costs is attributable to the increased specialization of markets. Different groups of users have very different expectations of the machines they buy; the design of specialized machines must reflect that fact. But an even more significant factor in rising R & D costs is the rush to mass production resulting from the decision of micro manufacturers to target the home market. For, while the mass market can be enormously lucrative, it is also enormously fickle and most micro manufacturers have little experience in designing machines for the general public. The home buyer may know little or nothing about computers. What is good design for a scientist, accustomed to the forbidding array of dials and strange computer commands, may deter or completely discourage a first-time user. This is apparently what happened to Texas Instruments' model 99/4, intended for the ordinary family. The 99/4 flopped in the marketplace for a variety of reasons, among them its high price, but it is significant that the model's later reincarnation sported a redesigned keyboard as well as a lower price. Even that was not enough, however, and Texas Instruments announced its intention to stop producing home computers in late 1983. Development costs for Apple's Lisa, which appeared in January 1983 after many delays, are rumored to have been fifty million dollars. Most of the cost represents an effort not to make the machine more powerful but to make Lisa easier to operate. Company officials refuse even to speculate about the costs of developing the McIntosh, Apple's newest project, which was recently introduced. Clearly ease of use and the accessibility of the machine to the average person are of utmost importance for this young industry which aspires to invade every home by 1990.

Making micros easier to operate, or, in computer talk, more user-friendly, requires the skills of a variety of professionals working together toward the same goal. Some of these professionals include:

- ergonometricians
- information specialists
- product designers
- psychologists
- market researchers.

Ergonometricians are engineers who specialize in the man-machine interface; as computer engineers, their efforts are generally directed to the inside of the machine. The skills of information experts specializing in the small machines are necessary in this area as well. Product designers are needed to make the appearance of computers both attractive and practical; they also ensure that computer controls are intelligently organized. Of course, to accomplish these goals the product designer must call on the knowledge of others. A button which the engineer knows will receive heavy use must be put in the most convenient place; it must also be made of a durable material. Although at first glance the computer's outside appearance may seem to be a trivial consideration, it is vital both in selling the computer and in guaranteeing consumer satisfaction. Whether, for example, to house input/output devices, the CPU, and perhaps even certain peripherals, such as a printer, together or separately is a product design decision. Other design choices would include the color of the cabinet and its height and the location of strategic controls. In making such decisions and in testing their effect on consumers, psychologists' skills come into play. Sometimes even sociologists may have a role in this aspect of production. Of course, market researchers are important as well. It is they who garner, sift, and evaluate information about what the consumer wants and relay it to the producers and designers. They also test the acceptance of the product in the marketplace.

The manufacture of microcomputer hardware creates many interesting jobs for skilled professionals from various disciplines, yet actual production is done by traditional assembly-line methods: the complex tasks involved are broken down into simple ones which can be performed by the unskilled. *Almost any workers* in the world *can and do* find employment on such a rationalized assembly line. As the industry matures, however, the experience of other manufacturers tells us that robots will perform more and more of these unskilled tasks. If forecasts for micro production come true, robots must predominate on assembly lines in order to meet production schedules. The introduction of robots, in turn, will provide employment for roboticists, robot software designers, and others.

SOFTWARE. Just as the production of hardware is capital-intensive in this industry, relying heavily on huge investments in automating machinery, software production is labor-intensive. In fact, the production of software is essentially at the craft stage, requiring thousands of individuals whose high-level skills create unique products. While hardware can be produced virtually anywhere in the world, creating software demands a pool, or infrastructure, of trained and experienced programmers to turn the hardware into a useful product. Software can only come from developed countries which possess such an infrastructure, which takes years to develop. At present, only the United States and Canada have an experienced pool of programmers; even the Japanese are far behind, though, having become aware of their deficiencies, they are rushing to catch up.

To understand why software creation must be so labor-intensive, we might consider how a programmer operates. An experienced programmer working, say, in the data processing department of a large corporation with a familiar computer may produce several hundred lines of material per day, but that won't run on the computer. So the programmer must "debug" the several hundred lines, throwing out the "garbage" until there are probably only ten to fifteen usable lines left at the end of the day. Keeping in mind that many average programs run upwards of thousands of lines, it is easy to see why software creation takes so many worker hours and therefore costs so much. No assembly-line workers would think of a finished micro as their machine, yet programmers ordinarily do talk about *their* programs: such an attitude is akin to that found in handicraft. The industry will continue in this way at least in the near future, due to the structure of computers. Software has been characterized as "the greatest boon to cottage industry since the spinning wheel." While perhaps 150 companies are manufacturing hardware, literally thousands of individuals are writing software. In the foreseeable future, employment opportunities in programming will rise astronomically.

If we say that the computer business in general is booming, we must say that the software business is exploding. In 1980, American software companies reported sales of $13.14 billion. By 1984, forecasters predict sales of $33.8 billion. Today, companies supplying mainframe and minicomputer users account for most of that total. But business for micro software producers, too, shot up from almost nothing in 1977 to $2 billion in 1982. Business applications account for most of that total. Although, as we have seen, the market for home computers is still minuscule, 1981 saw $250 million in sales of software specifically for use in home computers; projections are for $1 billion by 1985. In fact, according to International Data Corporation, a Framingham, Massachusetts, computer market research firm, sales of all forms of micro software have grown two and a half times in two years, a phenomenal 57 percent compounded annually. Currently 10¢ to 15¢ is spent on software for every dollar spent on hardware; that proportion should be more like 25¢ to 35¢ for every hardware dollar by 1985. Many compare the hardware/software situation to buying a stereo. After making the initial investment, people go on buying records year after year. As it now stands, the software industry is very much like the record business with hundreds of producers on small, private labels and many free-lancing for or employed directly by bigger companies. It had always been the pride of IBM that the company made its own chips and established industry standards with its own software. But not even dominant IBM, which controls 57 percent of the world computer market, can keep up with all the customized needs of its micro customers; IBM is now in the market for free-lance software.

The voracious public and business appetite for quality software has created an enormous demand for computer experts of every stripe, particularly in the more technical operating, language, and utility types of software. The specific names of the occupations may vary considerably, but among them will be found:

- software engineers
- systems designers
- systems engineers
- systems analysts
- information systems managers
- product engineers
- communication specialists
- software troubleshooters
- programmers.

Many of these occupations are highly technical, but in a high demand/low supply market, employers are more willing to provide on-the-job training. Salaries are high. Promotion can be very rapid; it depends little on scholastic background or school grades. Computer software specialists are rewarded for one thing only: achievement.

Becoming a computer expert is easier than one might think, but it does usually require some formal computer courses. In times of exceptionally high unemployment, when even MBAs and Ph.D.s sometimes go begging for jobs, Gary Kilbertus, a twenty-four-year-old university dropout, is now a systems engineer for Datapoint in Montréal. After graduating from a three-year data processing course at a local junior college, he entered a university data processing degree program but left after two years because he felt the university had little more to teach him. With no experience and no degree, he did have some slight difficulty finding his first job. He went to work for RCA as an operator, a low-level position, but departed two years later, now supervisor of the computer room, to work at Spar Aerospace as a junior programmer. Another two years later, when he had progressed to the position of intermediate programmer, Datapoint hired him as a systems engineer. One day soon Gary hopes to open his own consulting business. How important is school? Gary thinks school really only showed him the computer. He picked up absolutely everything else either on the job or on his own time. Nevertheless, in order to get a foot in the door, he advises taking at least a three-year computer course. And, he adds, pick up as many computer languages as you can.

Working for a large company on a salaried basis is one way to enter this high-tech industry, but not everyone is so technically oriented. The almost insatiable consumer demand for software has led many to follow alternate routes, trying to satisfy it. Small software firms are springing up everywhere; frequently the founders are teams of people joined together to exploit their radically different skills. Typically some members of the team are computer experts or professional software designers, while others are business people or knowledgeable in specific areas where the company hopes to specialize. Some companies create and package their own product; others, called software "publishers," act mainly as clearinghouses for free-lance producers.

Software is an almost ideal product for the entrepreneur with little money but lots of imagination and drive. Creating software doesn't take money; it takes brains. Because software creation is a "cottage industry," requiring long hours of development, it often appeals to those who have free time, such as teachers and students.

Because applications software must be "about" something, professionals from diverse areas, from business to biology, stockbroking to statistics, have much to contribute. Although spare-time entrepreneurship means working long days, it can represent the perfect combination of independence and the security of a steady job. Career changers or those who simply want to add a little spice to their working lives take note: the would-be software entrepreneur can start part time or full time, on a salary or as a free-lancer. Some firms specialize in certain areas, such as business or medical applications; others offer programs covering an unimaginable variety of topics.

Computer language programs and applications software are what most attract the small business and home buyer. Business financial planning and word processing packages, which can run into the hundreds of dollars, are industry best sellers now. But as the number of home buyers mounts, that preference will quickly change. Industry insiders say the greatest software need today is for simple, self-explanatory applications software. Yet, to the casual observer, packages already available for use in a home computer seem multitudinous. *The Book* for 1983 lists what appear to be hundreds of programs to run on your Apple. PC Clearinghouse Software Directory, published by P. C. Telemart, Inc., of Fairfax, Virginia, identifies 21,000 micro software packages from 2,912 different sources. Commercial applications programs cover a range of subjects the average person would never dream of, as the following small sampling shows:

MICRO CHESS: three levels of difficulty, for expert as well as beginner; thinks ahead as many as three moves. $20

COMMON BASIC PROGRAMS: seventy-six programs addressing problems in math, science and home budgeting. $15

AIRFLIGHT SIMULATION: practice take-offs and landings with full instrumentation. $10

ASTRONOMY: introduction to stars and twenty-seven constellations; gives celestial coordinates, annual motion of sun; quiz included. $32

SOLAR PAK I: aids design of solar homes and panels by computing amount of solar energy hitting a surface; plots solar heating and sun angle. $10

CALORIE COUNTER: considers your height, weight, sex and age, then estimates your average caloric requirement. $6.50
PERSONAL FINANCE: balances your checkbook, computes revolving charge account payments, calculates interest, payments and value for loans, etc. $10
TAX/INVESTMENT RECORD KEEPING: aid for organizing and recording data in a single, flexible system; keeps track of assets, liabilities, income and expenses. $70

Now that you've got the idea, think up one of your own . . . and you're in business. Hundreds of software companies really did start that way. As an eighteen-year-old Harvard student, William Gates founded Microsoft in 1975; the Bellevue, Washington, firm had 1981 sales in the $15 million area. Microsoft is famous for its version of BASIC, which has sold over 500,000 copies. Gates never did get back to Harvard. At twenty-seven, Daniel Fylstra cofounded Personal Software, producer of VisiCalc, a $200 financial forecasting program which has sold 200,000 copies and spawned dozens of imitators. His initial investment was $500; two years later, in 1980, sales were $4 million. It is worth underlining here, however, that as the buoyancy of this market becomes more and more evident, the heavyweight multinationals are moving into commercial software. Competition in this field will surely get tougher for the small companies. For those who prefer to test the waters before taking the plunge, Lifeboats Associates of New York is an example of a software "publisher" selling only programs written by free-lancers. Among its 200 packages are MicroSpell and the expense-account organizer T-Maker II. Royalties run from 15 to 25 percent.

What makes a good software writer? Aside from technical ability and experience, which are absolutely essential, the most important factor, according to Ellen Dahlstrom, a Bell Canada analyst in Montréal, is good communications skills. Ellen works for Systems Education now, but, in a fairly atypical Bell career pattern, has had experience in many groups within the computer division. Before obtaining her data processing training, Ellen completed a BA in sociology and psychology. What she learned at university has served her well, she says. In particular, her education taught her how to talk easily, a vastly underrated skill; how to be comfortable reading 600-page books; how to write effective reports; and, especially, how to interview people. Her ability to interview has helped her in many ways. To implement a self-development program aimed at improving her technical skills, she interviews the more experienced consultants and advisors among her coworkers, who help her select and evaluate her study material. In effect, she has made them her teachers. Her interviewing skills are also helpful in dealing with groups of potential users who often do not really know what they want the program to do. In fact, analyzing and structuring the complex process of understanding and communication that must take place before the creation of a program is only now becoming common practice among professionals.

Almost anything can be a salable subject for a software package. Managing Bowling League Data, for instance, is a program produced by Rainbow Computing which records the pertinent scoring data for leagues of up to forty teams. Sex Role helps you examine your nature, behavior, and attitudes in light of society's changing concept of sex roles, according to Creative Computing Software's advertising blurb. Educational topics are also fertile ground for microcomputer software. Not only do many people buy home computers as much for the education of their children as for themselves, but schools of all kinds are also buying micros in ever-increasing numbers. As usual, the universal lament among teachers and students alike is the dearth of good-quality software. The Minnesota Educational Computing Consortium's Wrong Note, an imaginative program to teach sight reading of musical scores, is unfortunately the exception rather than the rule in educational programming. Computer education in the schools is too vast a topic to discuss adequately here. This topic will be discussed at greater length in the section "Resource Sector," on p. 71. For the moment, it is enough to say that schools will initially use micros—and thus need software—in teaching the handicapped and in all types of remedial work requiring constant repetition. The need for teachers alert to the potentialities of small computers and for software producers alert to the needs of students is great. The growth potential for good educational software, a rare commodity, is staggering.

SUPPORT SECTOR

The role of the support sector is often obscured by the excitement over a new technology; discussions of a new enterprise often neglect the many jobs generated in this sector. Support personnel are those who administer the day-to-day affairs and guide the development of a company, yet who are not directly involved in the design or creation of products. Their role does not appear to be very glamorous. Especially as a young industry matures, however, their contribution is absolutely necessary if the industry is to thrive. When an innovative, transformative product like microcomputers establishes itself and begins to expand into the mass market, competition suddenly increases. So, too, do the problems of production, distribution, and service. The support sector exists to minimize the problems while insuring the survival and prosperity of the product and thus the industry. The people in the support sector come from many different areas of expertise and display many kinds of skills.

In the early days of the microcomputer creating a viable product was enough. Customers beat a path to the door of the manufacturer rather than the reverse. All that has changed now, of course, and the small entrepreneurial companies which until recently were the norm in both hardware and software manufacturing must expand to meet the demands of a new market or sink into oblivion. They must also meet the challenge of the recent entry of the major corporations, with their many financial resources and wealth of marketing experience, into what was previously private turf.

Both large and small companies, but especially the smaller ones, must devise ways to raise capital in order to increase production, cut costs, and improve efficiency so as to stay competitive. Their task is made more difficult by an economy in which high interest rates and tight money have become a way of life. Investors are reluctant to take risks these days, so unless firms can demonstrate that their management teams are guided by sound financial principles, they will not get any money. Right now is the opportune moment for business professionals of diverse types to enter the burgeoning microcomputer industry. In particular, the following professionals are needed:

- managers
- administrators
- accountants
- lawyers
- financial experts
- loan/venture capital specialists

and others with managerial skills. Naturally previous experience in the computer industry or an interest in micros would be an asset.

In addition to administrators and financial experts, advertising and marketing specialists are in great demand. Marketers help an individual company plan its sales strategy and decide how best to reach its designated market. Advertising relies on teams of people, from photographers to account managers, writers to conceptualizers, to make the public aware of a specific company's product. Both groups are made up of people with exceptional communications skills. For job seekers at least a nodding acquaintance with micros is an asset. At this point in the development of the microcomputer industry, marketing and advertising specialists have especially delicate tasks. Aside from making their product known, they must also educate the consumer in the uses of the home computer. They must also deal with the public's fear of computers, which is far from dead. IBM'S 1983 personal computer ads featuring a Charlie Chaplin look-alike were award winners precisely because they so smoothly combined all these tasks into one sucessful marketing campaign.

Micros are generally marketed through the retail distribution system. Until recently, the 1,500 micro retail outlets in the United States, led by Radio Shack with 655 stores and Computerland with 222, handled the bulk of sales from all manufacturers. But with the phenomenal growth in the home computer market, that distribution system has become woefully inadequate. The expansion of the retail sector holds vast possibilities for the alert entrepreneur. The average markup on a micro is well over 30 percent, and as we have seen, the sale of the machine is only the beginning. The cost of most peripherals is extra; software purchase is an ongoing activity like record buying. Opening a retail micro outlet looks like an excellent proposition for the eighties and nineties. Any apparent gold mine, however, attracts intense competition.

Now that the mainframe heavyweights have finally deigned to notice the micro market by manufacturing hardware, they have also put their extensive marketing and service experience to work. IBM and Xerox market their machines through their own business centers, at Sears, Roebuck and Montgomery Ward, as well as through mail order solicitation. These multinationals should easily capture a sizable market share relatively quickly because they have the resources for massive advertising campaigns and such customer support techniques as 800-number hot lines and fast service. The nation's most prosperous retailers have also gotten into the distribution act. R. H. Macy is now selling Atari and Texas Instruments computers with plans to add the lines of other manufacturers. The Winstock stores and the Dayton-Hudson chain carry micros. Canada's Eaton chain also sells micros. There is no doubt that by 1985 at the latest, large department stores in every major city will stock home computers, just as they now handle video games.

Selling micros is far more complicated than retailing video games, though, as anyone who wants to market them successfully must be well aware. In the first place, the public is largely uninformed about micros. For the immediate future, any successful retailer has to be as committed to selling the *concept* of micros as to selling the actual machines. That means, among other things, spending more time with browsers than does the traditional retailer and supplying, free or otherwise, a good selection of printed supporting material.

Selling the concept also means providing a dedicated, informed, and communicative staff. Students in particular should take advantage of this opportunity to get into the business: good micro salespeople are worth their weight in gold; many of them are teenagers. An intelligent entrepreneur will make every effort to hire those who are both sincerely enthusiastic about the micro concept and willing and able to communicate that interest to others. The entrepreneur must undertake to train selected sales personnel. Salespeople are truly the key to making this entire industry a success or a failure. Too often in the past the computer store salesman, typically a home hobbyist himself, confused and intimidated the potential customer with a barrage of technical jargon. With intensifying competition, such retailers will go the way of the dodo bird. In the development of a new industry, word of mouth is responsible for many sales. New style retailers, who are interested in helping their clients learn about micros and make the right selection for their needs, will be rewarded by those same customers with good recommendations.

The opportunities in retail distribution are enormous. Throughout the 1980's and into the 1990's, individual stores, chains, franchises, and computer sections of large department stores should flourish. In addition, many astute entrepreneurs are already developing lucrative sidelines. The selection of hardware and peripherals is difficult enough for the first-time buyer, but the selection of appropriate software is a nightmare for newcomers. There are hundreds of available programs, ranging in quality from terrible to fantastic, offered by numerous unknown companies with some of the best ones founded only yesterday. So supplying consulting services, particularly in software, is a logical and lucrative spin-off for the retailer, who has the most immediate potential access to customers. Some retailers are also imaginative enough to provide a staff person to customize software to the needs of the client, be it for business or home use. In fact, independent sales outfits which do nothing but package micros and peripherals with appropriate software for doctors, lawyers, accountants, or even farmers have already entered the scene. Retail distribution in its many forms is, without exaggeration, the hottest aspect of the micro industry today.

Effective distribution of computer hardware relies upon repair and service personnel for repeat business. This is heartening news for the job seeker: micro technicians will be in great demand, for there are few such technicians today. Just as the first wave of communications technology created a need for television and radio repairmen, now the second wave brings the micro (wo)man into existence. Technical institutes and high schools would do well to add the appropriate courses to their curricula. Micro technicians can also help in the proliferation of the product while generating extra income for themselves, for many people will acquire home computers by buying secondhand or reconditioned machines.

RESOURCE SECTOR

The resource sector is potentially the largest segment of the industry. When initial demand for hardware in business and in the home has been satisfied, consumers will still be avid for information. Creating, providing, and organizing information is the business of the resource sector. Throughout the eighties and nineties job seekers with varied skills will find employment here in increasing numbers as the microcomputer establishes itself. Most jobs in this sector require communication and research abilities. Curiously enough, print literacy will be a *sine qua non* for jobholders in this sector.

The jobs of the resource sector fall roughly into two categories: educational and informational. The educational area is responsible for extending computer literacy through every means possible, including the public school system. Coincidentally this segment plays a support role by aiding advertisers and marketers in their task of alerting the public to the potentialities of micros. This endeavor will create salaried and entrepreneurial jobs by the score. The informational area is responsible for the establishment and maintenance of data bases and information services. Already a large industry, data base networks have generated countless positions in the last few years; they will continue to do so in the future.

EDUCATION. Colleges and universities have long had access to computer systems; almost from the beginning, these institutions, engaging in computer-aided research, produced nearly all the personnel qualified to work in the industry. Now, as cheap micros make computers accessible to everyone, even the youngest students can join in. Today's parents, convinced that computer literacy must become universal, are pressuring school boards to buy micros for their children's classrooms. Initially supported by federal aid, the financially strapped boards are now turning to the manufacturers for help. Steven Jobs, young multimillionaire cofounder of Apple Computer Corporation in Cupertino, California, wants to give an Apple to every public school in the country, a total of 80,000 units worth $200 million retail, in return for tax credits. Recently, he persuaded two legislators to introduce bills in Congress permitting computer manufacturers tax credits for donations to elementary and secondary schools, as is already allowed for donations to colleges and universities. The State of California is granting such tax credits for a limited period. Through its Kids Can't Wait program Apple is donating fully equipped Apple II Es, including software discount coupons to 95 percent of the 9,200 eligible schools. At $2,400 each, the computers' retail value is $20 million, though with tax credits the real cost to the company is about $1 million. That figure, however, does not account for time volunteered by company employees and dealers, for Apple had to make sure that at least one person in every school knew how to operate the machine. In many cases local dealers helped train the teachers.

The micro revolution in the classroom poses several obvious questions, among them: Who will educate the teachers? Public-school teachers are notoriously shy of the machines which over-enthusiastic futurologists, who appear to know little about education, have predicted will take over their jobs. Computer-aided instruction (CAI) has great potential in the classroom, but it is highly unlikely that the human presence in the education of the young will ever become obsolete. The teachers' teachers will perhaps come initially from manufacturers, although with the sharp rise in demand this could only be a short-term solution. Alert entrepreneurs will, no doubt, step in to fill the gap.

Administrators of public schools with newly acquired micros will search frantically for instructors and good-quality software. Consultants offering comprehensive services should experience a boom extending into the early nineties. In fact, individuals or groups able to teach courses introducing micros and programming, whether to schoolteachers or to the general public, are already desperately needed. Adult education administrators in universities, colleges, and high schools as well as in private organizations would like to make more such offerings but cannot because of the shortage of qualified teachers.

In addition, many extracurricular groups devoted to computer education are springing up, bringing with them even more employment possibilities. Some of these groups are staffed by volunteers; many are not. Computer clubs in schools are generally volunteer. Summer camp organizers have already jumped on the computer bandwagon. These groups are not volunteer. Even Club

Med, experimenting with computers as a diversion for the children of vacationers at a spa in Mexico, found the idea so wildly successful with the adults that soon singles will mingle over micros at its resorts all over the world. In 1983 Universal Systems for Education, Inc., opened its first Personal Aid to Learning (PAL) Center in Denver, Colorado. A private business, the PAL Center uses Apple computers and its own software to teach reading to elementary school students, always, of course, under the guidance of a certified teacher. PAL Centers will soon be franchised nationally. Imaginative entrepreneurs will discover many more money-making ideas in this area.

Like good computer teachers, good-quality educational software is in short supply. So far, schools are generally using programs to direct drills and reviews. Hundreds of programs for micros popular in the schools, such as the Apple II Plus, Commodore's PET, and Radio Shack's TRS-80, are quizzing, prodding, and grading students much in the manner of teachers at the turn of the century. Cosmetic touches, such as a *Wow!* or a *Hooray!* flashed on the screen for a correct answer, cannot compensate for the robot-like quality of this type of software, which is, alas, all too common. Imaginative software takes money and time to develop.

An encouraging sign, however, is the thoughtfulness displayed by those working on the problem. Joe Major, a systems engineer for IBM, is currently devoting his time to creating an educational software system he calls MUSE. According to Joe, the software dilemma, particularly for schools, has a number of facets. First, individual programs are not practical for general use. Users spend more time learning the ins and outs of each particular program than they do learning the material. What is needed is a systematic approach with a set of programs all responding to the same indices. Second, in communicating with a program, users always discover the personality of its author. In itself, this is not a bad thing, but as things now stand, it is invariably unlucky for the user because most programmers are, in Joe's words, "intellectual sadists," determined to prove either themselves or the computer smarter than the user. What is needed is a humanistic, positive attitude to programming. Third, educational programming displays a one-sidedness in its method of communication similar to that of the authoritarian classroom. The teacher is always right and allows students no freedom of choice in solving problems: the conventional way is the right way. What is needed is a method for learning through discovery, so real learning can take place.

MUSE, still incomplete and experimental, is Joe Major's attempt to provide a humanistic, creative teaching system which can cover many topics. To date, the multilingual system, communicating simultaneously in French, English, German, and Italian, teaches a number of subjects, including cities of North America, modern painters, modern architecture, mathematical concepts, and chemical concepts. The system is multiprocedural, meaning one can learn many different things from it at the same time; at the very least, it teaches four languages along with whatever topic is chosen. The object of the system is to make ideas interesting through games; the user can choose either strategies or puzzles. Each subject or topic within a subject is independent of all the others; each is presented through concise descriptions. In the math section, for example, there are eighty topics, all at the high-school level and all self-contained.

Experimenting with MUSE has made its developer acutely aware of the hurdles faced by educational software designers, both in the long term and in the short term. Overcoming these hurdles will require the services of many people, most of them not computer professionals; as always, software creation is labor-intensive. One of the greatest difficulties is the lack of a theoretical model of how learning takes place. As a long-term project, the complex task of developing such a model is for the psychological researcher. Also in the long term, logicians and educational experts must figure out how best to structure the learning experience for the student. Both theoretical projects point to the need for extensive research. In the short term, MUSE, now run on a graphics terminal of the IBM 4300 computer, is far too expensive for schools, though people with computer training could adapt it for use on a personal computer. For the time being the selection of topics on MUSE remains small because Joe writes most of the material himself. To be an efficient learning tool, the system and others like it must be more comprehensive.

For the future, Joe envisages large numbers of people with specialized knowledge, "knowledge-bank creators," he calls them, who will develop and write the programs for the various subjects of study. The skills necessary for this job have little to do with the computer. What's more, computer literacy is no big deal, Joe says. In fact, it's like driving a rental car: a few fumbles with unfamiliar dials and switches and off you go. For example, a French teacher who helped Joe had never seen a computer before but learned how to input data in just one day. What is difficult is selecting, organizing, and writing up the data for the knowledge bank: the entry must be economical with words, extremely precise, yet still interesting to the user. This means that print literacy will continue to be a most necessary skill. In the future, individuals will receive an even greater proportion of their written information from computers, so the quality of that writing must be even better than it is today. The employment prospects for knowledge-bank creators are wide open.

The traditional print media also have an important function in educating the public about computers. Already a number of computer magazines have sprung up to supply both technical and general information to their readers. Such journals as *Interface Age*, *Microcomputing*, and *The Byte* have become a familiar sight on newsstands and in drugstores. These magazines inform their readers about the latest in hardware and microprocessor design and, to some extent, provide a forum for unbiased debate about various offerings of the industry, much as *Road and Track* does for automobiles. These magazines also aid in the gradual dissemination of computer literacy while providing employment to many. Books about computers have also become a kind of subindustry in the publishing world. Some of these books appear on the lists of mainstream publishers, but many authors, taking a hint from the fast-paced industry they write about, have set up houses of their own—such as Dilithium Press or Osborne, recently acquired by McGraw-Hill—devoted exclusively to computer-related subjects. In yet another development, newsletters for micro industry insiders, such as *Infoworld*, fill a real need. It is obvious from these few examples that the computer, far from making the print media obsolete, as has often been predicted, in reality contributes to their growth.

Most prominent among those who work for the various print media are, of course, writers. Writers able to explain the intricacies of the mighty micro in a language free of the omnipresent computer jargon are scarce. Even more such writers will be needed in the future to supply material for books and for all industry-generated publishing. Both hardware and software come with printed manuals for the customer's use, but they are often written in that strange language, technogab. Technical writers able to translate manuals written in technogab into a humanly comprehensible language are also prize commodities. Not to be forgotten are foreign-language translators who are proficient in computer vocabularies and understand the rudiments of the technology. The export market is a large one; frequently manufacturers display chauvinism by expecting everyone to read and understand English. This is not always the case even in North America, as the example of Quebec makes clear. Although it would seem obvious that translations are mandatory even for the home market, the current absence of foreign-language word processing programs, to name just one example, makes one doubt this is clear to manufacturers.

INFORMATION. As we have seen, the personal computer, appropriately programmed, can do almost anything around the house. The micro, however, has other, potentially more far-reaching capabilities. The home terminal, equipped with a modem connecting it to telephone lines, can communicate with any similarly outfitted machine anywhere in the world. More significantly, the home terminal can tap into enormous data banks, giving the average person, for a fee, quick access to information on every topic under the sun. Already 1,450 data bases now exist in the United States, and they are growing exponentially; dozens of new ones are created every month. Today's industry serves mainly the business, professional, and research communities. Telerate, a New York company providing financial information to banks, brokerage houses, and investment firms, for example, has increased its revenues elevenfold since 1978. At the moment, the company is installing its screen terminals at the rate of 2,000 a month. Information utilities range from traditional providers like Dow Jones, Dun and Bradstreet, and *The New York Times* to such unlikely competitors as *Reader's Digest*. TRW, one of the biggest American space and defense contractors, moved into the information business in 1969. It is now the largest information operation in the

United States, maintaining files on 86 million customers and processing 250,000 credit reports a day.

A few consumer-information services now have the lion's share of the still-tiny home market. As the home micro industry grows, however, so, too, will the market for these services. Joining the young industry today guarantees a solid occupational future in the world of tomorrow. The Source, with a central mainframe in McLean, Virginia, is the first and largest of the emerging "information utilities" designed to provide public access to computer data bases. In exchange for a $100 subscription fee to The Source, the home computer owner gains access to more than 40 separate data bases and 1,400 specific information and communication services. Via The Source, subscribers can, among other things, play an electronic game with a partner 3,000 miles away, monitor legislation pending before Congress, make a reservation for a trip, review the latest best sellers, or shop for over 30,000 discounted items. CompuServe Information Services of Columbus, Ohio, makes a similar offering. Fees are low. Computer time on CompuServe, for example, costs $22.50 an hour, but the rate is discounted to $5 an hour on weekends, holidays, and at night; The Source charges $15 an hour for prime time, $4.25 after six, and $2.75 after midnight. As the market grows, other information utilities set up shop. Access, an acronym for American Consumer Communications and Entertainment Systems, for instance, started nationwide service in 1981.

MicroNet of Columbus, Ohio, offers a number of consumer services to the home-computer owner. Home hobbyists can use MicroNet's computer to store programs too large for the small memory capacity of their own machines. MicroNet also provides, again for a small fee, bug-free software of every variety which would take the home hobbyist months or even years to develop. Through electronic mail, subscribers can leave messages for any other subscriber or, through the electronic bulletin board, for groups of individuals. MicroNet also runs a variety of clubs which enable micro owners to swap programs; Heathkit owners have one channel, Apple aficionados, another. MicroNet's engineers have devised, among many other ingenious items, a CB radio simulator, complete with interference, atmospherics, and so on.

Accessing MicroNet or any other information service, while requiring highly sophisticated telecommunications technology, is relatively simple for the user. A subscriber in New York, for example, dials up the information service computer, which may be located anywhere in the country, through the modem which is connected to the normal telephone lines; the New Yorker pays only for the local telephone call plus computer time. Through microwave relay or a private telephone system owned by the information service, the subscriber connects to the central computer. If the subscriber wishes to hook up to, say, the CB radio simulator, the call is routed through another microwave relay to other subscribers also connecting to the simulator. Subscribers may be located anywhere in the country yet pay only for a local telephone call. Schematically the hookup looks like this:

Services and information do not come into existence all by themselves. Data-base networks garner their information in a variety of ways, depending upon the subject. A review of best sellers, for example, might be prepared by an employee of The Source, or an in-house researcher might collect and summarize reviews appearing in the popular press. Alternatively a literary journal might, for a fee, agree to provide the service to The Source. However each specific informational item is gathered, it provides employment for those we may call content experts. Such experts need not be intimately familiar with computers, but they must know how to write up the material in a precise and accurate way. Communication skills, especially print literacy of the highest order, are, as ever, of supreme importance.

Obviously, in order to keep costs low, a data-base service pays for as few informational or service items as possible. Having the supplier pay for the privilege of using the network is the ideal situation. Career Network, carried by The Source, is a nationwide multiple listing of job opportunities and résumés compiled by select recruiting firms in the United States and supported by a membership of firms in other English-speaking countries. Computer Search International of Baltimore puts the material in computer-understandable form and provides key-word indices so that users can quickly sort through the information on the basis of a number of different specifications, such as salary range, geo-

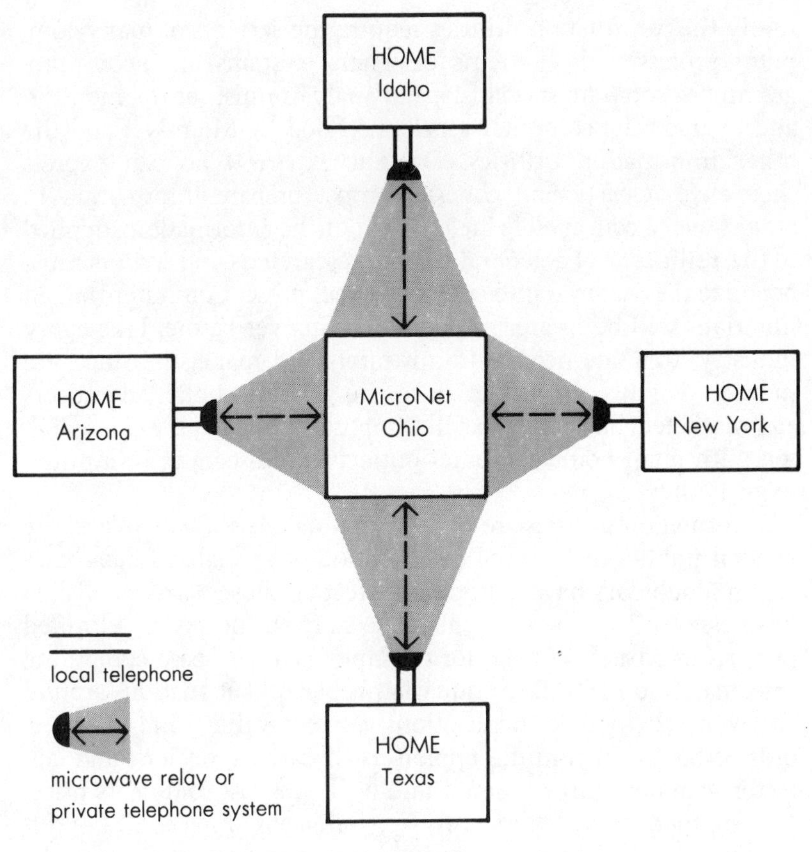

FIGURE 4.3

graphic location, job experience, and education. The Baltimore company then supplies the material to The Source. In such an arrangement everybody is happy. The Source has another service with which to entice subscribers. The subscriber gets a job. The participating recruiting firms get commissions. The middleman computer firm gets payment for services rendered and an ongoing source of revenue.

It is impossible to discuss all the positions generated by the information utilities and related companies. But perhaps a brief sketch touching on only a few areas will help point the way. Naturally the information utilities require the services of many computer professionals. Systems designers, systems engineers, programmers, software specialists, and analysts must set up, monitor, and upgrade the computer services offered by MicroNet and the other information utilities. Content experts who can express themselves clearly and concisely must prepare information in almost every conceivable area. Much of the information supplied to the utilities is of a general nature: researchers and writers must organize the material into compact, exact prose. Computer-trained librarians will be in greater demand than ever before. Like every industry, this one needs administrators and managers. Since the industry is new and unknown to the general public, advertisers and marketers have an especially important role to play. Close liaison with retail home-computer outlets would seem to be a prime requisite here.

Information utilities are not the only data bases available to the general public. In fact, the proliferation of specialized data bases will undoubtedly have a profound effect on those with established businesses or those who are about to start businesses on a limited or part-time basis. Utopia, for example, is a data base containing information on 150,000 antiquarian books sold at auctions around the world. It includes descriptions, current values, and sales records of books. In minutes appraisers, librarians, dealers, and collectors can determine when a rare book was last sold or its price. Having such information easily accessible could make all the difference to a small firm or an individual who cannot afford to make costly mistakes. More established firms might find that consistent use of such a data base increases profits. Horse breeders or owners can avoid nightmarish errors by checking references in Horse. This file records breeding and race information for more than one million thoroughbreds in North America since 1925.

Other general-reference data bases of interest to the public are:

MAGAZINE INDEX: listing more than 500,000 articles, editorials, short stories, recipes, and reviews from 370 magazines,ranging from the *Harvard Business Review* to *Playboy*;

INFORMATION BANK I & II: containing more than two million abstracts of news and editorials published in *The New York*

Times since 1969, along with current information from 10 other newspapers and 50 magazines;
NEXUS: providing full texts, rather than the more conventional abstracts, of five years of articles from the *Washington Post* and selected major magazines, newsletters, and wire services.

The computerization of such information serves to supply users in minutes with what it used to take researchers hours, days, or even months to find in conventional libraries. Certainly writers and those investigating almost any imaginable topic for whatever purpose will be enormously aided by such data bases.

Further, it is not generally recognized how valuable such information is. The primary product of consulting firms, for instance, whether they specialize in politics or mining or education, is information. Although much data is frequently available for almost nothing in newspapers, journals, and magazines, the information is extremely disorganized. In the past, organized information was only accessible to those with a good deal of time and the requisite library skills. Data bases, in organizing information, make it more useful. The existence of data bases should contribute to a marked increase in the number of information-producing organizations, such as consulting and research firms, and to a much improved informational product.

The future of the data-base industry is already assured. Even though the number of home micros is still small, the general public will initially benefit from data bases through a trickle-down effect. Intermediaries will do the actual telecommunicating, usually for a fee. For example, public-library patrons will receive bibliographic information through trained librarians. People who belong to a civic, social, or political organization will find they can access data through a terminal in the office. Data bases will aid enormously in the creation of thousands of informational jobs which will come into existence throughout the 1980's and 1990's.

CHAPTER FIVE

TELEVISION COMES OF AGE

Television, by any measure, is the supreme example of a successful technology. The spread of television technology in the United States was almost instantaneous. In 1945 there were 6,000 television sets in American homes; by 1950 that figure had already risen to 15.5 million. Television is also a powerfully transformative technology; its impact is immediate and profound. During the 1950's, the three American television broadcast networks became an integral part of the American way of life. After a mere twenty years, television had completely altered the collective consciousness of the nation. It has had the same effect throughout the world, whether the technology was adopted early or late. By now, just about everybody in North America has at least one set; the norm is more like two or three. In fact, 81.5 million or 98 percent of American households have at least one television set, more than have telephones or indoor plumbing. Everybody watches television; network news commentators are among the most trusted people on the continent. Watching television is, by far, America's favorite leisure activity.

But while the national average of hours spent in front of the tube rises every year, so do complaints about the poor quality of programming. "Intellectual pap," "schlock," and "a cultural wasteland" are a few of the kinder cuts. The sad fact is that North American network television, trying to be all things to all people, succeeds in pleasing very few. Americans, no longer the video innocents they once were, have been demanding radical changes for a number of years. Finally, thanks to innovative video technologies, the changes are coming. As a result, television is nearing the end of its babyhood and is about to enter its adolescence. Like any adolescence, it will be a time of experimentation and rebellion, of alternating highs and lows. It will be a difficult period but certainly an exciting one. During this transitional era, what is now almost a closed industry will offer plenty of opportunity for those eager to display their talents, as television realizes its potential. For there is little doubt about it: television is finally taking off.

The first indication of the momentous changes about to take place in television occurred in 1977, when, for the first time, the networks' share of the total television audience failed to increase. Little noticed at the time save by a few terrified network executives, that event has since turned into a trend. In the last six years, network audience share has dropped 16 percent, though networks still maintain a massive 77 percent of the total television audience. A survey by the Association of National Advertisers indicates the network audience will continue to fall, bottoming out at 70 percent by 1990 or before. The splintering of the television audience, once the virtual prisoner of the broadcast networks, results from inaugural adoptions of alternate video technologies. Videocassette recorders, video disks, cable and satellite hookups, and other promising innovations make possible an enormous variety of programs, all competing for the home viewer's attention. When the adoption of alternate technologies becomes more widespread, as it is sure to do, the fragmentation of the market will accelerate. Television as a medium will continue to grow, but the direction of its growth will be quite different. Television will go the way of radio.

When television first came in, radio, losing its network audience, seemed headed for oblivion. Yet radio is today a dynamic medium because its appeal was redirected to local and more and more specialized markets. Even record players, tape recorders, and proliferating CB channels have not diminished its impact. In fact, the resurrection of radio was in large part responsible for making the recording industry the giant it is today. Although the appearance of a new technology may temporarily stagger an established industry, the various media are not necessarily competitors and often help each other's growth. Sometimes new technologies, while creating new industries, may stimulate old ones.

Like network radio of thirty years ago, network television is about to lose its monopoly control over the American audience. It will undoubtedly continue to play a major role in television programming for quite some time to come, but specialized and local programming are clearly growth areas. In the next few years, the home television set will become less and less merely a passive transmitter of broadcast programming. It will, instead, become a true home entertainment center, providing a wide range of programming practically tailored to the individual viewer's desires.

Eventually the changing role of television will enlarge the entire entertainment industry, creating thousands of jobs and exciting employment opportunities. While the television market fragments, producing more specialized programming becomes economical as soon as a target market, however small, emerges. The recent spread of cable, a relatively old technology, points the way. Not three networks but tens of channels are hungry for programming. Once the dust settles and markets stabilize, the motion-picture industry could well be the beneficiary. Due to a voracious video market, the 1990's will be the Golden Age of Entertainment.

How is all this possible? Can television really rise from a sea of mediocrity and begin to fascinate you once more? Can there really be new jobs in this familiar industry? The best way to find out is to examine some alternate technologies already on the market as well as others still in the experimental stage.

ALTERNATE VIDEO TECHNOLOGIES

Today's broadcast network television is a twelve-billion-dollar-a-year industry whose profits regularly return 50 to 60 percent on investments. Network television earns every penny of its astronomical profits from advertising. Commercial television is, in fact, a strange hybrid; it is an unnatural combination of two dissimilar and unrelated industries: consumer products and home entertainment. Such a marriage, though often awkward, has made commercial television one of the most gigantically successful mass media marketing strategies ever devised. But that hybrid is about to split asunder with the advent of cable and its offspring, pay television. The reign of network television as sole monarch of the airwaves is just about over, at least in the form in which it has dominated the television screen for the past thirty years.

Network television has developed into a successful business by utilizing the public airwaves to hawk its advertisers' products.The Federal Communications Commission (FCC) regulates the industry, but, like the phone company, commercial television has enjoyed a virtual monopoly on what is, after all, common property. Network television, hardly a model for the free enterprise system, is now facing competition for the first time. Network executives were right to be scared back in 1977 for, as a result of proliferating video technologies, one of America's most successful businesses is about to join the real world. The rules of the game have not changed. The FCC still severely limits the number of stations and licensees who are to have access to the public airwaves. At last count, there were 775 commercial television stations in the United States, the vast majority affiliated with one of the three major networks. But clever entrepreneurs have found a way around the rules. Cable television uses not the public airwaves but wire, installed at great expense, to transmit its product to you.

CABLE TELEVISION

Broadcast television is by no means a perfect technology. Normally a conventional television station receives distant television signals through an antenna; the station then broadcasts the signals to the surrounding community via the airwaves. After traveling horizontally through several miles of atmosphere to get to the home, the broadcast signals are often weakened. Obstructed by hills, tall buildings, inclement weather conditions, and the like, the airwave signals frequently distort both radio and video reception. The television set suffers from "ghosts" or "snow," while hissing and crackling noises are intermittent accompaniments to the sound. In addition channel capacity is small, for the range of signals is limited by the very curvature of the earth's surface; it is not possible to transmit a conventional broadcast signal much beyond a fifty-mile radius.

Right from the beginning, however, enthusiastic television technicians and electronics buffs set to work solving the problem of poor reception due to obstruction or distance. In Lansford, a town about sixty-five miles from Philadelphia's broadcast signal, Robert J. Taylor had difficulty selling television sets because of poor reception. Business improved in 1949, after he began installing rooftop antennas on the homes of his customers, but he continued to mull over the technicalities. A few years later, Taylor came up with a better idea. Conjointly with the Panther Valley Television Company, he built a tall master antenna atop a nearby mountain, strung coaxial cable wires on poles to the town of Lansford in the valley below, and fed signals to paying clients. The public was obviously eager for the best video possible: the first Lansford cable subscriber paid a hefty $125 installation fee and a further $3 a month for better reception.

All across the continent alert entrepreneurs had the same idea. Canada, then as now, had an insatiable appetite for American programming. In 1952, E. R. Jarmain of London, Ontario, devised a special rhombic antenna in order to receive signals from distant Cleveland. By 1959, he had entered into partnership with Famous Players Corporation to form the London Cable Television Service; already in 1962, it had 10,000 subscribers. Montréal, Palm Springs, and Vancouver repeated the story. Bigger and better antennas, coupled with an experimental technology, resulted in a new means of transmission which could circumvent the defects of broadcast television. That, for the moment, was enough. In this way, the first cable television systems were born.

Cable's eventual role was implicit in the technology from the beginning. It is a relatively simple technology. Antennas, one for each broadcast channel, are mounted on a tower. The antennas receive television signals over the airwaves or, in the case of more distant transmissions, via microwaves. Then, through special receivers, the antennas feed the signals into a small building called the headend. After processing, the signals can remain in the plant for future transmission or be relayed directly to the subscribers' sets through coaxial cable hookups.

The coaxial cable trunk line, strung aboveground on poles or run underground, carries the signals from the plant throughout the cable system's licensed range. Feeder lines running off the trunk lines carry the signals to within 150 feet of every subscriber's house. When a new subscriber joins the system, a drop from the feeder line connects directly to the home television set.

The coaxial cable itself consists of a small inner conductor of copper wire, an insulator of polyethylene foam, and a surrounding outer conductor of braided copper wire or an aluminum sheath. Between ¼ and ¾ of an inch thick, the whole is covered by a relatively non-corrosive plastic outer sheathing. Since the conductors prevent current from radiating off the wire, coaxial cable is a powerful transmitter; its capacity is far greater than that of conventional broadcast television. Each cable can carry from twenty-eight to thirty-five television channels, the entire spectrum of AM/FM radio, and large amounts of non-visual data.

While originating a method for improving upon the defects of broadcast television, the inventors of cable technology inadvertently stumbled on the means to far surpass it. In those early days, however, since television itself was yet only an emerging technology, organizing the industry, producing broadcast programming, and coping with the effects of such a powerful new medium were priorities. The potential of cable systems to utilize their far greater transmission capacity to produce original programming in competition with the broadcast networks was thus not even realized until recently.

Canada, for example, is one of the most wired countries in the world relative to its population. Yet Canada produces little original cable programming. More than 50 percent of the nation's households receive cable. The penetration rate in large cities is even higher; cable reaches 86 percent of Vancouver's households and 70 percent of Toronto's. The reason Canadians took to cable so early and so enthusiastically is not hard to discover. With a small population widely scattered over the gigantic territory of the second-largest country in the world, Canada is a natural for cable. The presence of many broadcast channels so tantalizingly near, just over the border in the United States, makes cable seem merely a way to enhance network reception. In fact, Canada's 1968 Broadcasting Act mentions cable technology only as a means of extending and improving reception quality; its potential role as a source of original programming is not even considered. Nor is Canada alone in its lack of vision.

The main activity of almost all cable systems on the North American continent is retransmitting ad-supported programming originated by others. Recently, however, the public clamor for something more interesting to watch has resulted in yet a new twist: pay television. Pay television and its sometime sidekick, pay-per-view television, are not new technologies. Rather, they are a first attempt to completely break commercial broadcast television's consumer products paradigm. Pay television, a nascent industry in the United States and Canada, has begun to realize the potential of cable to provide a much wider variety of entertainment than is presently offered on commercial networks. By charging a reasonable fee above the price for basic cable service, cable operators can offer pay television and the home subscriber the chance to see the current best in film, drama, dance, and music—all commercial free.

Even as this relatively old technology starts to fulfill its promise, however, yet another innovation has emerged which, though now a partner of cable, will eventually become a much stronger competitor. The communications satellite can provide far more channels than cable and an infinite variety of programs from all over the world.

SATELLITES

Approximately 300 working satellites hover in geostationary orbit 22,300 miles over the equator. Although about 200 are for military purposes, the remaining 100 are there for civilians. Some collect data for weather forecasts; some serve as sophisticated communications satellites, for telephones, for business, and, increasingly, for television. Indeed, satellite technology provides so versatile a communications conduit that, throughout its short history, one terrestrial industry after another has found a use for it.

The International Satellite Organization (Intelsat), a 109-nation international consortium in business since 1965, is a global system implementing all applications of the technology; both the United States and Canada are member nations. Intelsat now has more than fourteen working satellites in space; it provides all international television transmission and two thirds of all intercontinental telephone services. More than twenty countries lease Intelsat services for domestic communications. Early communications satellites were used almost exclusively for telephone trunking. Early Bird, the first Intelsat satellite, connected Europe and North America via one sixty-conduit, long-distance trunk. As satellite technology improved, though, not only did the number of telephone circuits increase and the charge per circuit decrease, but other industries saw ways of utilizing the technology as well.

Recently corporations have begun using satellite business teleconferencing services. Teleconferencing can take the place of much expensive business travel via New York Telephone's Picturephone Meeting Service. For a $600 fee, a company can set up a ninety-minute coast-to-coast conference. No airfares are involved, no hotel rooms, no entertainment—just business. Within a decade or so, teleconferencing equipment will be standard in all major corporate headquarters; businesses will buy their own earth stations. Businesses have long used satellites for transmitting data, but now satellites can handle all communication modes. All-digital satellites are just now coming into service. Through wideband channels, these satellites transmit all material in digital form almost instantaneously, regardless of the nature of the original message—whether voice, data, facsimile, video, or industrial control.

Such a versatile technology seems like science fiction, but yesterday's dream of the future is today's reality. The communications satellite is a tightly packed container of electronic equipment, launched into geostationary orbit by rocket or space shuttle. Once in orbit, the satellite maintains a constant position in the sky, for, by traveling at 6,800 miles per hour, its speed synchronizes with that of the earth's rotation. Inside the satellite, transponders, combination transmitters and receivers, do the communication work. The transponders receive a signal from one ground terminal and, seconds later, have already transmitted to another ground terminal hundreds of miles from the first. The ground terminal, or earth station, consists of a nine-to-twelve-foot-in-diameter dish antenna, angled toward the appropriate satellite, as well as electronic equipment which processes, amplifies, and relays the signals onward.

The United States now has three domestic communications satellite systems: Westar, owned by Western Union; Comstar, owned by AT&T; and Satcom, the property of RCA. Westar and Comstar are primarily for telephone and business use, while Satcom provides the bulk of television programming now available through domestic satellite. More domestic satellite systems are planned. Most American communications satellites use the C-band (6/4 GHz) frequency for transmissions. Since this frequency is much higher than that of broadcast television, the signal can travel far greater distances, literally from one end of the earth to the other.

Using satellite technology, many conventional commercial television stations have become "superstations" with a nationwide broadcasting range. Ted Turner's Atlanta-based WTBS, for example, beams its signals from a satellite to affiliated earth stations across the nation which, in turn, relay them to the audience over the air or by cable. Satellite technology will shortly revolutionize the entire television industry. In 1981, eleven communications satellites offered 60 channels' worth of programming to the superstations. Considering expected improvements in satellite technology, it is well within the realm of possibility that up to 350 channels will be broadcasting coast-to-coast as early as 1985. Television manufacturers clearly anticipate increased channel capacity: RCA is already producing a high-tech television set able to receive 127 channels.

The public at large is also becoming involved. A small television receive-only (TVRO) earth station costs between $2,000 and $5,000; more than 30,000 TVROs are privately owned. Popular in hotels and large apartment complexes, these earth stations are one-way receivers, pulling in cable and pay television signals for free, as well as prereleased original programming, sent by networks to affiliated stations for later scheduling. Due to the ambiguous wording of the Communications Act, it is at the moment still legal to set up a private earth station without a license, as long as the programs are not shown for profit. In Canada, where the issue is just as hot, the Department of Communications issued a directive indicating that hotel and bar owners or others using satellite programming to attract customers could expect prosecution. Within months the department reversed this decision in the face of vigorous public protest. In self-defense, some programmers scramble their transmissions to guard against piracy; others have a wait-and-see attitude: earth stations are, after all, relatively expensive.

The issue of private earth stations is not about to go away, however; if anything, the problem is becoming more complex with the development of yet another innovative technology. It is now possible to buy a small dish antenna for the residential rooftop for under $1,000. The era of the Direct-to-Home Broadcast Satellite (DBS) has arrived. Satellites, formerly helpful partners of the cable and broadcast industries, suddenly appear as menacingly competent competitors. Bypassing the clumsy and expensive wire of the cable companies, these three-feet-in-diameter dishes can grab any television program directly out of the sky for you at home. All this is possible because transponders broadcasting in the higher Ku-band (14/12 GHz) frequencies now have commercial space to lease. The Ku-band transmissions are so powerful that much smaller dishes are efficient receivers. Canada, whose Anik A-1 was the first domestic satellite system in the world, now leases space on its three sophisticated Anik-C satellites to American pay television companies; applications already far exceed the existing capacity of the three in operation. The National Aeronautics and Space Administration (NASA) contemplates developing Ka-band transmissions (30/20 GHz). Within the next twenty years, the DBS–home-dish-antenna combination may well become the norm, replacing cable altogether.

DBS, in fact, was set to begin commercial operations in late 1983 when United Satellite Communications, Inc., financed by General Instrument Corporation and Prudential Insurance, begins service with five channels transmitted by the Canadian Anik-C satellite. Inter-American Satellite Television Inc., a New Jersey-based group, also intended to start a five-channel DBS service in late 1983. In addition, Satellite Television Corporation, financed by Comsat, planned to offer five DBS channels by 1984. All are aiming at areas, primarily rural and remote, not already wired for cable; there are about 30 million homes in such areas in the United States. The short-term economic future of these companies is doubtful, however. Costs to subscribers are high: $35 to $40 a month or a $400 installation fee and $15 a month. The logistics of servicing a clientele spread out over a large geographical area is another difficulty. When initial marketing and programming problems have been overcome, though, commercial DBS promises much.

For, besides its efficiency, DBS technology has another vital point in its favor: it delivers a sharper, clearer picture. Since large video screens or panels are the next generation of technology proposed by television manufacturers, a clearer picture is an important feature. Currently, conventional sets use 525 scanning lines to make up the television picture; when the image appears on the wall-sized panel screens, the lines become visible, resulting in a blurry, fuzzy picture. High Definition Television (HDTV), an experimental system now in the prototype stage at the CBS Technology Center in Stamford, Connecticut, increases the number of scanning lines to 1,100, greatly improving the quality of the picture. HDTV, however, requires a wider bandwidth on the electromagnetic spectrum than that of conventional broadcast transmissions. The wider the bandwidth, the greater its capacity to transmit information. DBS, it just so happens, makes use of a portion of the electromagnetic spectrum easily divisible into the broadband channels HDTV needs. HDTV and DBS, together supplying services consumers already want, are strong contenders as the television technologies of tomorrow. There are, however, a few snags.

Even a technology as marvelous as DBS, so close to giving us all hundreds of television channels to choose from, does have technical limitations. For one thing, the amount of space available for geostationary orbit is strictly finite. Since the launching of *Sputnik* in 1957, over 3,000 satellites have been launched into orbit. Satellites have a limited life-span, and a mounting number of old derelicts clutter up space and drift out of their assigned orbits. According to the National Oceanic and Atmospheric Administration's (NOAA) National Earth Satellite Service, a possible collision between a derelict and an active satellite is an ever-increasing danger. Also, satellites operating on the same radio-television frequency must be two hundred miles apart to avoid electrical interference. A recently inaugurated NOAA program facilitates operations by shepherding the active satellites from one slot to another, but the slots, at least in the Western Hemisphere, are becoming scarce. By 1990, the limited space may be exhausted. Finding alternate slots over other countries is proving to be a political hot potato. The list of nations using satellite technology lengthens constantly. Third World countries also depend upon communications satellites for educational programming and weather forecasts; these nations are fighting for slots as well. Colombia, for example, a country directly on the equator, claims the 22,300 miles of outer space directly above its land mass. Colombia fears, as do many other underdeveloped nations, that otherwise, richer countries, particularly the United States, will hog all the slots in the Western Hemisphere.

With the spread of cable and pay television, improvements in satellite technology, full-channel television sets, DBS, and HDTV, obviously the television industry, as it has existed up until now, is in for some major changes. Suddenly we realize how young a technology television really is, though more than half the present population has never known life without it. The complacent years, when viewers passively accepted whatever the networks presented, are over. For there are yet more innovative television technologies already delivering the viewer from bondage to the networks.

MORE ALTERNATE TECHNOLOGIES

More than 60 percent of the cable systems now operating in the United States offer only twelve channels; the hundreds of channels promised by satellite technology will not be forthcoming for some time to come. Yet home viewers, impatient for greater choice, have already demonstrated they are not content to spin the television dial in the hopes that some programmer, somewhere, is presenting something pleasing. In mounting numbers, home viewers are selecting their programming in retail outlets or even creating their own. The videocassette recorder (VCR) industry is booming because consumers are eager for choice.

VCR owners can watch whatever they want, whenever they want. For as little as $300 to $500, viewers can reinvent the television set. VCRs will record one program while the viewer watches another; more sophisticated programmable models will record many shows over a three-week period. Most machines have a remote control option, and top-of-the-line models feature special effects as well: frame advance and freeze frame; visual scan for rolling tape back and forth while watching the picture; variable-speed slow and fast motion; and now, Dolby stereo. The video camera, hand held and simple to operate, can make producers of us all. Video cameras use no film. Nothing needs developing. Anyone can aim the camera and watch the action, as it occurs, on the television set. The technology is perfect for recording events traditionally left to the home-movie buffs; it also makes creating fairly professional productions a breeze. The most sophisticated cameras on the market have innumerable functions until recently available only on conventional movie cameras: automatic focusing; editing capability; a character generator for superimposing letters and numbers on the tape; electric and variable-speed zoom lenses; and special settings which shoot 35 mm negatives and then show the photographs on the video screen.

All videocassette recorders are made in Japan. RCA is an American company, but its VCRs are made by Matsushita, the Japanese electrical giant; Zenith VCRs are made by Sony. Recently prices have been dropping even faster than usual because the Japanese, who don't often make mistakes in consumer electronics, have produced more than the export market can absorb. A considerably more advanced 1982 Betamax costs about the same as did the 1977 version, while costs for practically everything else have almost doubled in the same five-year period.

Videocassette recorders come in two basic formats: Beta and VHS. Beta format brands, such as Sony, Sanyo, Sears, Toshiba, and Zenith, claim to offer a slightly better picture, but only a video technician could tell the difference; Beta brands now have about 30 percent of the market. VHS format brands, such as Akai, Canon, Fisher, GE, Hitachi, JVC, and RCA, offer a longer recording time, a maximum of eight hours as compared to five hours on Beta; VHS format brands account for 70 percent of the market. With the appropriate accessories, both can record as well as play back prerecorded material.

The first videocassette recorder, Sony's Betamax, appeared in 1975. At first demand was sluggish: consumers purchased 160,000 sets in 1977. By 1981, however, over 3 million VCRs had been sold, 1.5 million in that year alone. For the first time, industry sales topped the $1 billion mark. With demand snowballing, forecasters predict that 1984 will see the purchase of 10 million sets. Eventually the videocassette recorder will duplicate the success of color television, once a luxury item but now a fixture in the vast majority of American homes. Sales of VCRs also stimulate other new industries. In 1980, retailers sold 20 million blank tapes at prices ranging from $10 to $20 each, racking up $232 million in sales; analysts predict that demand will grow so rapidly that by 1984 retailers will sell 75 million tapes, giving this subindustry $1 billion in sales.

Rising interest in alternate programming has also created a rapidly expanding market, shared with video disk players, for video club retailers, who sell and rent prerecorded material, usually movies. Video disk players, even cheaper than VCRs, cannot record off the air, but they have the advantage of much better picture quality. Prerecorded movies for use on home VCRs or video disk players cost from fifty to eighty dollars to purchase; the movie producer gets a seven-dollar royalty fee for each. In 1981, between four and five million such films were sold. The real jump in business, though, is in the rental of films through video clubs. Rental fees are inexpensive, from two to ten dollars a day; the retailer can up volume by also renting the necessary video equipment. A copyright problem is in the making here, however, for, like authors who receive no royalties when their books are photocopied, the movie mogul gets no part of a rental fee. And according to surveys, 52.8 percent of all VCR owners record films rented from video clubs or right off the air for their own libraries. The most likely solution to the problem is a special surcharge on blank tape sales as a kind of blanket royalty fee; so far, no one can agree how the money could be equitably distributed to movie producers.

All television technologies show enormous promise, but change will take time. For the moment, the greatest growth is in producing and installing equipment or operating retail outlets. By the 1990's, however, programming and its myriad supporting industries will offer incredible opportunities to those who begin preparing themselves now.

THE TELEVISION INDUSTRY: JOBS

In the future, the television screen, now holding pride of place in the living rooms, bedrooms, and kitchens of the nation, will continue its dominance. As a result of the booming electronics industry and its steady stream of micro marvels, television sets may become smaller, flatter, lighter, and more portable; they will certainly become even more numerous than at present. Over the short term, at least, the television set will look much as it always has. For all the proliferating video technologies, adding a VCR or a video disk player is about the only improvement to the basic apparatus a home viewer can make. But however much the old set may look the same, it will never be the same again.

The most radical changes resulting from the introduction of

emerging video technologies to the mass market are occurring in the television industry itself. Once the exclusive domain of the giant broadcasting corporations, their programmers, and their suppliers, the television industry is expanding at a rate not seen since the beginning days of the original technology. In the *technical sector*, cable systems are now a well entrenched and growing part of the industry; the importance of satellites and affiliated technologies increases every year. Even NASA, criticized by some for wasting taxpayers' money on useless pyrotechnics like the moon walk, has, in a manner of speaking, become part of the industry. With its space shuttle, NASA has for the first time put its technical achievements on a paying basis. Rapidly expanding industries, aside from creating job openings for the technically oriented, also, of course, multiply positions in the *support sector*. Administrators, organizers, managers, marketers, and their cohorts are vital in any industry; they ensure not only the success of start-up operations but also their continued prosperity. The television industry is no exception. As cable and its partners in technology make possible an array of channels never before even imagined, the *programming sector* must broaden its scope. Providing choice to the home viewer is the real purpose of these technologies. The promise of alternate entertainment is the reason for their exceptionally warm welcome from the public. A durable success depends almost solely on programming. In the long term, the most exciting prospects as well as the heaviest responsibilities lie in the programming sector. There should be many challenging opportunities for the creative and the talented throughout the 1990's.

THE TECHNICAL SECTOR

CABLE. Although a mere 30 percent of American households are now wired for cable, the expansion of the industry, once it got started, has been striking. In 1976, cable households in the United States numbered about 10 million; by 1982, the figure had more than doubled to over 20 million. Displaying a parallel rise in subscription rates, pay television entered a negligible 470,000 homes in 1976, but had captured a market of over 12 million households in 1982. The cable growth rate has been so spectacular that speculations about the future of the industry vary widely. Predictions for cable penetration rates range from Nielsen's conservative 50 percent increase by 1990 to an optimistic 90 percent. That 12 million subscribers were willing to fork over an extra ten to twenty dollars a month for pay television indicates that the outlook for that industry is equally bright. In fact, the predictions of a 300 percent increase in pay television subscriptions are quite possibly accurate. In these days of high-priced entertainment, pay television can bring into the home first-run movies, Broadway plays, and sporting events unavailable elsewhere for less than the cost of a ticket to the ballet, the opera, or a college football game. Naturally an increase in pay television subscriptions of that magnitude would necessitate the wiring of many more homes and therefore encourage the spread of cable.

In 1982, a little under 4,700 cable systems were operating in the United States, serving more than 12,000 communities. About 300,000 homes are wired for cable every month. Among cable operators, 40 percent are choosing to offer the pay option. Some cable operators are even giving away the basic service, normally priced at seven to ten dollars per month, in order to attract customers to the more lucrative pay options. On the average, the American cable client also buys 1.5 pay services along with the basic system. At present, the cable television industry's revenues run about two billion dollars a year.

In Canada, cable is a mature industry; already high penetration rates are not expected to increase by much more than 3 or 4 percent a year. With such a well-developed industry, one would expect the Canadians to be making big profits. The truth is, though, that Canadian cable is a troubled industry, hard hit by the national economy's tailspin. From 1976 to 1980, while industry revenues rose by 78.6 percent, expenses, fueled by high interest rates, rose by 87.2 percent. Profits are fading fast. In 1980, the industry was in debt to the tune of $131 million; that figure could easily become $1 billion by 1985.

The outlook for cable in Canada is not all that bad, however, for the lucrative pay television services have just begun operations. At present, the majority of cable systems in larger cities, some installed twenty years ago, can offer only twelve channels; 30 percent of the rural population still receives only one television channel or none at all. As a result, many citizens have erected illicit dish antennas in their yards or on their roofs; some 750 pirate earth stations have done likewise. In contravention of international agreements, both groups are receiving signals from the United States. In part to remedy this situation, the Canadian Radio-Television and Telecommunications Commission (CRTC) licensed six pay television services, which began operations in 1983. Expectations are that pay television will generate $200 million in revenues during the first five years, clearing net profits of $20 million, though the demise of C Channel after a mere six months of operation may indicate such estimates are too optimistic.

Despite the growth potential for American cable, there are problems in the United States as well. The 60 percent of American cable systems now limited to twelve-channel capacity need modernizing. Construction and installation costs over a ten-year period could amount to $5 billion. Merely laying cable underground costs $7,000 to $10,000 a mile. Nevertheless, companies in some cities, such as Valparaiso, Florida, have expanded from twelve to thirty-six channels and offer their subscribers a hurricane alert warning system and two-way home security services. As always, an industry's problems mean jobs for us. In addition, since cable wire eventually deteriorates due to age or extremities of weather, upkeep and replacement activities are ongoing sources of employment.

In 1982, according to the National Television Association, 40,000 people worked in the American cable industry, a 60 percent increase over 1975. Although some cities, such as Palm Springs, California, are already 99 percent wired, many, including Denver, Milwaukee, Boston, New Orleans, Atlanta, and Washington, D.C., are less than 20 percent wired. In those cities and many others, cable systems are growing at a rapid rate. With expansion, of course, come many employment opportunities. Start-up operations require technical personnel and often cable companies search for technical talent locally. In addition to jobs directly connected to cable systems' operations, there are other employment possibilities.

Anixter Brothers, Inc., for example, helps put up the works for cable. A major distributor of all types of wire, Anixter sells its products to phone and utility companies, mining concerns, and the cable television industry from its sixty warehouses in the

United States, Canada, and Europe. Employing 2,600 people, Anixter is the largest American distributor of cable and paraphernalia for the industry. Besides coaxial cable, the company markets connectors, splicing materials, aerial and underground construction materials, convertors, headend electrical distribution equipment, and satellite receiving antennas. Anixter has it all; this list of products gives an indication of the potential of the manufacturing sector of the cable industry. As cable grows, gadget makers and clever entrepreneurs will discover many areas where their products meet a real need.

Burnup & Sims, employing 5,000 people throughout the nation, is a leading manufacturer and installer of wire. This company, responsible for over 20 percent of all the cable now in place across the United States, has laid over 100,000 miles of the stuff. A diversified concern, Burnup & Sims is sinking ever greater investments into the video communications industry and it recently acquired Gardiner Communications, manufacturers of satellite dish antennas. While Burnup & Sims made about a million dollars from the cable industry in 1980, 1982 revenues were more on the order of ten million.

General Instrument Corporation of New York is the market leader in headend equipment and black-box converters; it also installs coaxial cable. In 1981, this company laid 100,000 miles of wire and distributed 3.5 million converters. Although General Instrument got the contract to wire Boston, it's not putting all its eggs into the wiring basket. In 1982, it ventured into the DBS–earth-station industry, forming, jointly with All-Star Satellite Network and Pop Satellite, a new company known as United Satellite Television. United planned to make its own three-feet-in-diameter earth stations available for under $1,000.

Throughout the 1980's and 1990's, with the continuing expansion of the cable-satellite industry and allied businesses, the demand for engineers, computer experts, and other technical professionals will remain high. There is an especially bright future for engineers of all types, electrical, electronic, industrial, mechanical, and civil, in the video communications industry. In the cable industry, electrical and electronic engineers are particularly important, for they design and oversee the manufacture as well as the testing and distribution of electronic cable equipment.

Along with draftsmen, engineers map out the cable system's electronic layout, tracing the flow from the antennas, through the headend into the plant, and thence to the homes of subscribers. It is now a common practice in the United States for licensing municipalities to require cable operators to propose the construction of separate coaxial cable networks for general-purpose broadband communications among local institutions and commercial centers. Such a network might provide the means for hospitals to exchange diagnostic information or for banks to exchange data on branch accounting. Engineers must create these systems as well.

Every cable system must have an engineering department to oversee day-to-day operations and maintenance. Engineering and electronics experts in this department are also responsible for the operation of the headend and the plant. Once the basic cable system is laid out, they do the electronic mapping of newly subscribing hotels, motels, and high rises; a dwelling of more than four units requires a complicated and precise electronic layout of its own. Some cable systems use microwave transmission to and from distant stations. The microwave engineer is in charge of maintaining optimal conditions. Sometimes individual cable systems of distant localities join together to form networks; this will probably happen more often in the future. In the event of networking, the skills of the microwave engineer are essential; thus these professionals should have no difficulty in finding employment in the video communications industry.

SATELLITES. The 100-channel television systems promised by cable operators can only be made possible by satellite technology. The future of the video communications industry lies in the sky. In the past few years, the satellite industry, trying to satisfy the demands for communications conduits from cable and pay television operators, business, the military, and telephone and telecommunications companies, has rushed to expand services as rapidly as possible. While satellite communications is already a $2-billion-a-year industry, growth rates, past, present, and anticipated, are phenomenal. Intelsat, the international satellite organization, provides a good example. With twenty-two separate systems now operating worldwide, Intelsat's 1981 revenues of $213 million were but a small portion of the global industry's total. Since the late 1960's, Intelsat has experienced a growth rate of better than

20 percent a year; it confidently expects a growth rate of better than 15 percent a year until the year 2000 and perhaps thereafter. For at least the next twenty years, qualified technical personnel should find green pastures in the satellite industry.

Another related industry where qualified technical people will find employment is the national and international space program. Government-run space programs and the private satellite industry are interconnected. The industry depends upon the rockets and shuttles supplied by space programs for launching its satellites into orbit. In the future, satellite repair may also fall within the realm of the space industry. Governments are no longer content to pour vast sums of money into prestigious and prodigiously expensive space flights for purely scientific purposes. While it is unlikely that NASA, for example, will ever pay its own way, elected officials are clamoring for commercial ventures that will defray at least part of the cost of the space flights. At the moment, the most attractive application for the United States government's space shuttle as well as for the entries of other governments is satellite launching.

The communications satellite industry is now growing so rapidly that the supply of available rockets and space shuttles is vastly outstripped by demand. Nevertheless, the several organizations now capable of meeting the demand, including NASA, the French government, and even nascent private initiatives, are in hot competition for the business. NASA's reusable space shuttle, the *Columbia,* successfully launched two communications satellites, one American and one Canadian, during its first commercial flight in November 1982. The American satellite, owned by Satellite Business Systems of McLean, Virginia, handles long-distance calls, teleconferences, and computer data; Telesat Canada's Anik-C blankets most of that country with television, telephone, and business communications. Both satellite owners paid the relatively modest fee of nine million dollars to NASA for the launch service. This price compares favorably with the twenty-five million dollars Telesat Canada paid NASA to launch its Anik-D satellite by Delta rocket. Naturally, any satellite company would prefer to use the space shuttle at those prices, but such bargains will be increasingly hard to find. Due to budget cuts at NASA, the number of space flights has been pared. The five NASA shuttles originally funded by Congress are now four and the number of launches has been correspondingly modified. As the number of launches decreases, the cost per launch will increase. Thus, for example, launching the next three sixteen-channel Anik-Cs in 1983 and 1984 will cost Telesat Canada nine million dollars each, while for the 1985 launch of a twenty-four-channel Anik-D, the fee will jump to around nineteen million dollars. Even that doubled price is a bargain. After 1986, the cost will probably exceed fifty million dollars. NASA has already booked sixty-two domestic and fifty foreign satellite launches. Some commercial users, however, are looking elsewhere for a ride into space.

Another popular vehicle is the Ariane, a non-reusable rocket developed by ten nations with the French government as the chief financial backer and controller. Arianespace Corporation, launching its rockets from French Guiana, expects to garner about 30 percent of the international satellite business in the coming decade. Originally planning to build five rockets a year, Arianespace will probably offer eight launches a year by 1985 in order to take care of its $350 million worth of backlogged orders. Arianespace's price is definitely competitive: about $25 to $30 million per launch. In addition, through its French banking connections, Arianespace offers financing at 9 percent, payable after the launch, while NASA demands cash, beginning almost three years before the launch. In these days of high interest rates, financial sweeteners can mean savings to the customer of between three and five million dollars per mission. The Ariane route is risky, though. In September 1982, its first operational flight failed when the third stage lost power and its two communications satellites fell into the Atlantic Ocean. By mid-1983, however, an Ariane rocket succeeded in carrying satellites into space, thus challenging the American dominance of the lucrative space launch market for the first time.

Other countries also have plans to cash in on the commercialization of space. Japan's fledgling space program, the National Space and Development Agency (NASDA), is slowly learning how to build its own commercial space launchers, using cast-off American Thor and Delta rocket technology. If all goes well with its new H-1 rocket, NASDA should be able to launch 1,200-pound satellites by 1987. The next generation of communications satellites, scheduled to go into service by the mid-1980's, will weigh over four tons, however. So NASDA's only hope of catching up technologically lies in collaborating with NASA. NASA is willing, for a price. To get the needed money, NASDA has been courting the powerful Japanese Ministry of International Trade and Industry, trying to get space technology designated as an industry targeted for special attention and funding. NASA, no doubt, will wish it luck.

Even private entrepreneurs, smelling money, are trying to get started in the launching business. Space Transportation Systems of Princeton, New Jersey, the first private launch company, now has a greater financial base, for it was recently acquired by Federal Express Corporation of Memphis, Tennessee, whose overnight package delivery service generates $1 billion a year in revenues. Federal Express subsequently announced the formation of Fedex Space Transportation Company, a joint venture with Martin Marietta Corporation, a military rocket supplier. Fedex proposes to use Titan rockets to put two satellites into orbit for Intelsat in 1986. Intelsat will most probably award the contract to carry one satellite to NASA for a reported 100 million dollars and will give the other to either Arianespace or Fedex for about the same sum. If Fedex wins the contract, it will become the first successful privately owned space launch system. If it loses, Federal Express nevertheless intends to get into other aspects of the enormously lucrative space market, by financing, insuring, and providing technical assistance for commercial users of space; creating a space-based electronic mail system; and developing its own fleet of satellites.

Fascinated by the immensity and challenge of space and lured by the thrilling accomplishments of space technology, many, especially young, people wish to make their careers in this field. The frantic activity in the satellite industry and the space program can only be good news for those with such a career path in mind. Now new opportunities are opening in these industries every day, and the future looks good.

Although times are generally tough these days for the aerospace industry, companies involved in satellites or the shuttle are prospering. As a result of space-related activity, in the next few years Canada's aerospace work force will increase by 1,000 people over the present 2,500. Spar Aerospace in Quebec, for example, employed an extra 600 high-technology workers when it manufactured the arm for the United States space shuttle *Columbia.* Opportunities in the United States are, of course, many times greater than in Canada. Some of the many professionals needed in space technology and related areas, such as the aircraft industry, are:

- engineers of all types, especially aeronautical, aerospace, electrical, electronic, mechanical, civil, and testing and instrumentation
- computer specialists of all types
- relevant technicians of all types
- designers
- physicists.

The success of NASA's commercial ventures may well benefit its purely scientific projects. In these days of high unemployment and budgetary constraints, it becomes increasingly difficult for elected representatives to justify funding such "useless" projects as the *Voyager* missions. Despite hard times, however, various polls indicate that the spectacular flights through the solar system have captured the imagination and won the approval of the American people, who, after all, are footing the bill. One highly unscientific poll was conducted in Columbus, Ohio, where subscribers to an experimental QUBE system can answer questions by punching key pads attached to their television sets. When asked if they thought *Voyager* missions justified their $400 million price tag, 88 percent of the respondents answered yes. If such grassroots support continues, and NASA can partially pay for itself through commercial flights, the result will certainly be job opportunities for pure scientists as well. Astronomers, biologists, theoretical physicists, and even chemists and medical researchers as well as technologists can aspire to the stars.

The national space program, as a result of its involvement with the satellite industry, is generating thousands of jobs; the communications satellite industry itself is also fertile territory for the qualified job seeker. Besides designing and manufacturing the satellites, communications satellite companies are supporting strong research and development (R & D) programs in the hopes of surpassing current technological limitations. Naturally R & D efforts mean jobs for technical personnel.

As mentioned before, one current headache for the industry as it strives to satisfy the insatiable public demand for more television, telecommunications, and data satellites is that the number of available satellite slots in space is limited. The solution will most assuredly be maximum utilization of each satellite. A variety of methods for maximizing a satellite's potential have been proposed. CBS and several other organizations are experimenting with bandwidth compression, a process by which satellite channel capacity can be doubled or even tripled. Thus fewer satellites could receive and transmit more information. Another method for maximizing the potential of satellites already in orbit involves transmission frequencies. Satellites now operate in the C- and Ku-band frequencies. With the projected increase in the number of transmissions, however, the capacity of the satellites now in service should be exhausted by 1990. NASA is presently developing a demonstration system of a new satellite operating on more powerful frequencies. Slated to appear in 1987, NASA's demonstration system uses the Ka-band (30/20 GHz) frequency range and includes several ground stations capable of receiving the system's signals. Since the Ka-band satellite uses narrower radio beams than those using C- or Ku-band frequencies, the satellite fits into a narrower orbiting slot. As a result, more satellites can be squeezed into the same space without fear of electrical interference. In addition, there are even those, especially at NASA, who envisage maximizing the potential of each orbital slot by using spacecraft in these positions instead of satellites. Dubbed "switchboards in the sky," these spacecraft could handle an immense amount of communications. But the technical hurdles, not to mention the financial ones, are formidable.

Another area in which new developments in the communications satellite industry have stimulated growth is earth station manufacture. Numerous companies have seen their stocks leap and their profits jump because of ventures into this budding market. The newer, cheaper dishes also open vistas of enormous profits in the mass market. Frost & Sullivan, a New York market research firm, projected, in 1978 dollars, that earth station purchases would increase from $47.6 million in 1979 to $164.5 million in 1988. The growth of the cable industry has accounted for many sales of receive-only stations that pick up signals from satellite programmers. Scientific Atlanta, Inc., has so far cornered about half this market with sales of about $265 million in 1981. Microdyne also markets a receive-only dish. Two-way earth stations, costing about $50,000 each, will soon grace the rooftops of office buildings across the continent, in response to another enormous and growing market. Network radio broadcasts are also transmitted by satellites. The number of earth stations used by radio broadcasters will probably exceed 5,000 by 1985.

The recent surge of interest in better and more varied television programming has resulted in the formation of potentially giant industries, cable and its probable successor, satellites. While the number of technical positions available in these industries will be large, even more jobs have yet to be considered. For the opportunities in what may be called the support, or organizational, sector of these industries will be multitudinous.

SUPPORT SECTOR AND RELATED INDUSTRIES

Finding a practical application for a technology and then creating a salable product are the first essential steps in establishing a new industry. Even technologies as exciting as cable and satellites, however, would never get off the drawing boards without the solid organizational efforts of many different types of people. Industries do not spring into existence all by themselves. A good technology does not guarantee a good business. From nothing, capable people, well versed in many disciplines, create a viable something. They conceive and organize the venture. The organizations they create must not only establish but also maintain and expand the business. The job is difficult and challenging. The role of the organizers is hardly glamorous; they are barely noticed unless their efforts falter or fail. Yet new industries would have no chance of survival without them. The new video technologies need such people in abundance and will for quite some time to come.

Think about it for a moment. Setting up a cable system is a complicated affair. Someone must sell the idea to investors or arrange for some kind of start-up financing. Operators of cable systems need licenses: they must bid against others eager for the same privileges. During this start-up period, cable operators need financial experts, venture capital consultants, accountants, and economists. Cable systems need lawyers and legal advice, since cable systems are regulated by federal law and sometimes by municipal and state law as well. Opportunities for consultants conversant with the technologies and the regulations are numerous. License in hand, cable operators need an organization, along with people skilled in identifying markets and the means to serve them. Cable, like any expanding industry, offers enormous opportunities in such support positions. Aside from the requisite technical experts, cable systems need personnel recruiters, administrators and managers, and, most of all, marketers.

At the moment, marketing is the single most important challenge confronting cable operators. In Canada, for example, cable systems, anticipating the arrival of pay television, are expanding their marketing departments. Rogers Cable System in Toronto has recently doubled its sales staff to sixty people. Rogers' marketing strategy even includes a door-to-door campaign to sell people on the new product, pay television, which could set subscribers back twenty-four to forty dollars a month. In fact, because of the advent of pay television, Donald E. Taylor of Rogers projects a 20 percent overall hiring increase in the industry.

The need to market the service certainly creates jobs, but while the cable industry is now growing, nobody knows for sure how cable will eventually support itself once it is completely established. There are a variety of methods by which cable could make its profits. Like the broadcast networks, cable in the United States will probably generate revenues through advertising, at least to a limited degree. American cable ad revenues increased from about $100 million in 1981 to about $250 million in 1982. Compared to the broadcast networks' ad profits of $6 billion, these figures are negligible. At this stage, advertisers are reluctant to sponsor cable programs. None of the cable networks has the range or mass audience of the broadcasters. Some cable operators are attempting to remedy this situation by banding together to form "interconnects" with bigger advertising markets. Hard, or physical, interconnects link numerous systems together by cable or microwave; in soft interconnects, a common organization coordinates advertising for the whole group of systems. In either case, small cable companies, serving as few as 30,000 people, can offer advertisers the benefits of a larger system; advertisers are more willing to invest in a twenty- or thirty-system network reaching a combined audience of perhaps 500,000. The interconnect approach works for Gill Cable of San Francisco, with thirty-two systems and an audience of 475,000. It works for a twenty-one-system network in New England with 500,000 subscribers and a thirty-seven-system network serving 144,000 clients in Idaho, Wyoming, Colorado, and North Dakota. In suburban Philadelphia, Harron Cable is spending $1.25 million to interconnect thirty systems in the area via microwave. The interconnect idea is really starting to catch on and may well generate not only new revenue for cable but also many more technical and support jobs.

Advertisers are also reluctant to commit money to cable because nobody knows how to measure the size of cable audiences. Network television ad marketing has turned into a slick and incredibly complicated business over the past forty years, spawning its own incomprehensible lingo and sophisticated methodologies. Nielsen ratings provide a consistent way to measure the broadcast audiences and thus the national television advertisers' potential market. When confronted with hundreds of channels from the twenty-five ad-supported cable networks now operating in the United States, however, Nielsen-type surveys break down. National advertisers are accustomed to the arcane world of Nielsen, but the advanced marketing methods used in broadcast don't work for cable. And lacking even minimal data, few television ad agencies know how to make cable work for big advertisers. Nielsen is, to be sure, devising experimental studies to measure cable audiences; advertisers and cable operators, though, can't wait. Marketers must come up with innovative methods to sell cable space to daring advertisers. Those who do will reap big rewards.

Among the few brave souls now testing the cable advertising waters is Anheuser-Busch, the beer maker. The company knows that 80 percent of the beer in the United States is bought by 20 percent of the population. Not surprisingly, most of that 20 percent are sports fans. So Anheuser-Busch advertises on the Entertainment and Sports Programming Network (ESPN), which currently has about twenty-five million subscribers. Instead of broadcasting to the masses, the company is narrowcasting through cable's specialized programming to that segment of the public which is most likely to respond to its product. To make sure its ad campaign is on target, the firm also supplements its national cable ads with spots on local cable systems, such as Sportschannel in New York City and Action TV in Philadelphia.

Marketing through the new video technologies is going to be very different for viewers and advertisers alike. Opportunities for clever entrepreneurs in this burgeoning industry are great, for as yet few understand how to exploit the advertising potential of the new reality. For now, at least, cable's limitations are also an attraction for advertisers. A major cable advertiser can play a dominant role impossible in broadcast television; fewer cosponsors and fewer local advertising cut-ins give a big advertiser far greater impact. Cable's inexpensive airtime, specialized audiences, and looser programming schedules allow immense flexibility in commercial length and target marketing unheard of in broadcast. On cable, even a regional retailer of a national sponsor can afford to make his own local pitch. In this way, a cable system expands opportunities for advertising firms located in its viewing area and familiar with local audiences.

Freed of the thirty- or sixty-second limit of broadcast television, commercials may break new ground. Some media consultants already speak knowingly of the "infomercial," "advertorials," and "infotainment"; they are already planning how to make even prerecorded videocassettes into vehicles for marketing products. All over the country, communications analysts and specialists are cracking their brains over cable—and coming up with bright ideas.

In Tallahassee, Florida, the research firm of Madison, Unger and Webb, Inc., developed a study, called Cablescan, of "the life-style characteristics" of that city's cable network audience. Tallahassee's Broadcast Center is marketing the study to interested parties. The Boeing Commercial Airplane Company launched its latest production, the 767, and jumped on the video bandwagon at the same time. Boeing made its own videotape of the ceremony and then distributed it by satellite to local television stations in six large American cities, the BBC, and an international news-gathering agency. Creative Associates of Louisville, Kentucky, one of the first production houses specifically for cable television, offers prepackaged ad campaigns for real-estate brokers, stockbrokers, travel agents, and others; the company works side by side with a cable system's ad sales force. Obviously, in this field, you create your own job. All it takes is talent and ideas.

As cable systems expand, salaried and entrepreneurial openings in television advertising will undoubtedly expand as well. As television programming becomes more local, so, too, will the television advertising industry, now located almost exclusively in the largest cities, such as New York and Los Angeles. Television advertising, whether local or national, employs production, technical, and support personnel. On the production side, enlarging the industry means more work for writers, directors, producers, actors, dancers, singers, and other entertainers. Sound and light technicians, stage managers, cameramen, and those with other video skills will also be needed. Support personnel will include account managers, researchers, salespeople, financial experts, and even psychologists. In addition, as television advertising becomes more common, even middle-sized companies may have to hire communications specialists and media directors, much as large corporations do already. Certainly graduates of recognized television, broadcasting, and communications schools and programs should be heartened by the public's ready acceptance of cable, for it almost assuredly means a brighter future for them.

By increasing the number of channels available to the home viewer, cable television and satellite technology fragment the video audience. Indications are, however, that cable also increases household viewing time: Nielsen estimates an extra three hours a week for homes with basic cable and an extra ten hours a week for homes with pay television. Advertisers will of course be quick to spot the potential here. Surprisingly perhaps, all the media, magazines, radio, even books and records, stand to gain from this extra viewing time.

Cable makes television behave more like radio or magazines because they frequently have audiences similar in demographic makeup. Appealing to the same audiences, the three are now becoming extensions of one another, reinforcing each other's impact, especially in advertising. Cable networks are often formed in conjunction with magazines. *Sports Illustrated* and Entertainment and Sports Programming Network (ESPN) operate in tandem, for example. The cable network gains from this association, for the magazine has a respected national identity, the authority of print, and vast stores of material potentially useful for television programming. The magazine also gains from its exposure to a wider audience and from the immediate impact of television. In this way, both media prosper and continue to provide employment to many creative people. Talented marketers invent new ways of reaching the public. *Sports Illustrated* and ESPN have an innovative merchandising package: the two may cooperate to send a golf pro, for example, to a company tourney. Kodak, Timex, Hilton Hotels, and Fuji Film are among the customers of these packages, priced from $300,000. Other magazines, such as *Popular Science, Outdoor Life, Ski, Golf,* and *Better Homes & Gardens,* are also dabbling in cable.

National advertisers have long relied on local radio spots to supplement and reinforce national broadcast television ad campaigns. Sometimes the introduction of cable reduces the ratings of local television stations, as viewers watch signals imported from other localities. Under these circumstances, area advertisers frequently turn to radio stations and newspapers. Unexpectedly cable can be a boon for these media.

Radio, in fact, may change drastically yet once again, this time as a result of satellite technology. For while attention is centered on cable television, network radio is on the rise once again. The new technology can create radio networks quickly and cheaply. Local ratings wars have, ironically, created a need for radio networks. Seeking even the slightest programming edge, hotly competitive local radio stations turned to syndicators or networks whose resources could provide performances or interviews by top-flight talent. Network radio accounted for $167.3 million in ad revenues in 1980, much more than cable's estimated $65.7 million. The long reign of the radio network kingpins, ABC, NBC, CBS, and Mutual, is threatened by the RKO Radio Network and the approximately twenty-five new networks that have appeared in the last several years. In addition, pay radio is soon to appear. Teamed up with record companies, pay radio will offer the cream of the music crop to its subscribers.

These days, radio is an exciting medium to work in, and it is likely to become more so in the future. Like cable, the radio industry needs production, technical, and support personnel. In addition, because the new networks are now offering increasingly specialized programming on a national basis, they need people who may never have imagined a career in radio. *The Wall Street Journal,* for example, broadcasts via satellite; researchers and financial analysts must provide the material. Literature and drama are returning to radio with *The National Radio Theater,* distributed via satellite to its own network of 300 stations, including National Public Radio affiliates, community and college stations, and even some commercial stations. News and information networks are exploding in popularity, as are sports and "life-style" features. The Leisure Market Network's thirteen stations target a mere twelve tiny communities, but whose denizens, its executives claim, belong to the top 5 percent of United States income groups. The many visitors to Aspen, Maui, Lake Tahoe, Lake Placid, and Vail, among other resorts, represent a potential audience of six million people for the new network. Never has radio looked so good.

Although many blame television for a decline in reading, the fact is that television helps keep the publishing industry viable. Television sells books, as ex-drug addict Alexander King was the first to discover during the 1950's, when exposure on *The Jack Paar Show* pushed his book right onto *The New York Times* Best Seller List. Since then, authors regularly appear on television to promote their books. In the mid-seventies, David Seltzer's film, *Omen,* became a paperback, which sold 3.5 million copies. This type of media interaction is now common. Cross-promoting *Star Wars* in many media—books, magazines, radio, and television—was enormously successful; *E.T.* and *Return of the Jedi* are but the latest examples of what cross-promoting can accomplish. The publishing industry has much to gain from the new video technologies. In fact, because of them, all the media will change and most assuredly prosper. Success is not guaranteed, however. Support personnel in a number of separate industries must work together for television's new reality to have maximum impact. Those aiming for the top in the media and media-related industries will need great dedication and ingenious ideas. The rewards are many, though, and the opportunities great.

PROGRAMMING'S "SOFT" JOBS

The metamorphosis of television naturally generates employment for technologists. In the burgeoning microcomputer industry, the frantic activity in the hardware end of the business pales into insignificance beside the informational, or software, side, which literally knows no bounds. If television's hardware side is experiencing explosive growth today, the real action tomorrow will be in software, known to television as programming. In response to the public's insatiable appetite for programming, hundreds of television channels will eventually be made available to the home viewer. The talented, the creative, and the just plain curious will have ample opportunity to experiment with video as the new technologies make television more accessible. Since the demand for video software will be virtually unlimited, many will create "soft" jobs for themselves in programming.

Many "soft" jobs are in the making right now for those who look in the right place. Even today, though the three major networks continue to dominate the home screen, an amazingly large number of non-traditional outlets are already screaming for programming, among them superstations, broadcast networks on cable, broadcast-network–backed ventures on cable, pay television, local cable stations, low-power stations, and even one adventurous offering bypassing all the above.

The superstations, brought directly to the cable operators by satellite, are just starting to turn a profit; from now on, they will give the major networks some real competition. In 1970, when Ted Turner bought the Atlanta-based WTBS, it was losing $600,000 a year. In a remarkable turnaround, WTBS, first of the superstations, now in 20.4 million homes, made $45 million in pretax profits in 1982. It features mainly movies, sports, and sitcom reruns. Encouraged by his success, Turner started the twenty-four-hour Cable News Network (CNN 1), currently received by 14 million subscribers, in 1980. Then in 1981 he created the Cable News Network Headline Service (CNN 2), sent to 1.8 million subscribers and also syndicated to 126 conventional stations. Although CNN 1 and CNN 2 are still losing money, Turner and the executives of his Turner Broadcasting System, Inc., recently announced plans to compete with the established networks on their own turf by starting a fourth broadcast network, Turner Network Television (TNT).

Sometimes known as "the Mouth of the South" to his many enemies, Turner has fascinating and innovative ideas for programming on his proposed network. Existing networks typically pay television production companies, usually located in California, $800,000 or more per episode for a prime-time entertainment show. As home viewers well know, this system has not worked well in the past few years. On the basis of short-term ratings samplings, nervous networks cancel programs just beginning to gain an audience and then scramble to fill in the time, leaving audiences disappointed and confused. Turner has no intention of buying programming. Instead, he plans to hand over time periods in TNT's nightly schedule to studios or producers, who can program whatever they wish for as long as they wish. TNT will make money by selling advertising; the programmers will make money by syndicating shows after their run on TNT. No one knows if this idea will actually work, but Turner has achieved far more than his rivals believed possible when he began his headlong rush to expand operations in 1976. Turner's success shows that tiny David can still battle the network Goliaths and win. In 1981, Turner Broadcasting had the greatest increase in revenues of the top 100 media companies, as tallied by *Advertising Age*. With his never-say-die attitude, Turner has also shown how powerful the new video can be; rumors have even had Turner forming a football league to play exclusively for television. Home viewers couldn't be more delighted.

Turner's remarkable superstations show that, even in these troubled economic times, subscriber television can become a gold mine. The proof of this is that the network giants are clamoring for a piece of the action and grabbing it any way they can. In 1970, the FCC barred the three major networks from owning cable systems, but in 1981, the FCC granted CBS a partial waiver of that prohibition. CBS is experimenting with High Definition Television (HDTV) via direct broadcast satellite (DBS), expecting, within five to ten years, to provide three channels. The FCC decided the experiment was in the public interest. Perhaps some of the commission members saw a CBS screening where the HDTV picture was almost three-dimensional in clarity with every color and fold in the material of the actors' garments visible. CBS's plans for its three HDTV channels, which will be transmitted nationwide, are quite specific. The first, an advertiser-supported channel, will be similar in content to CBS's regular network programming. The second and third, planned as pay channels, would offer educational, entertainment, and cultural programming, as well as information services. By experimenting with bandwidth compression, CBS hopes to free the DBS channels for other, unusual uses.

Now that CBS has won its waiver, the FCC has granted the network permission to own outright a cable system serving not more than 90,000 subscribers. CBS is also considering the acquisition of a number of smaller cable systems in as yet undetermined areas. ABC and NBC have submitted petitions to the FCC requesting waivers of the cable ownership restriction. FCC Chairman Mark Fowler has indicated it may be time for the commission to consider eliminating, or at least moderating, the multiple ownership rule in order to foster growth and competition within the industry.

The broadcast networks are, of course, involved in cable and satellite programming through subsidiaries and spin-off companies, but business has been far from good. The Satellite News Channel superstation, launched in 1982 at a cost of $150 million specifically to crush Turner's CNN, was owned by Group W Satellite Communications. Group W is a joint company formed by Westinghouse and ABC, which once vociferously opposed the whole idea of cable. Turner charges cable operators fifteen to twenty cents per month per subscriber to carry CNN. Group W was actually paying cable operators to carry its service: an initial seventy-five cents per subscriber and then ten cents per month. Of the competing group, Turner said, "I intend to dance on their grave." He declared war on Group W by starting up CNN 3 and planning a CNN 4, an all-day business news network. Just as he predicted, Turner won the battle, for in late 1983 SNC closed down. This further compounded Group W's woes, which began when their ambitious plan to found the Disney Channel fell through in 1982 with a most inglorious episode of corporate name-calling. Now on its own, Disney has committed more than $45 million to the first year of programing for its new channel.

ABC owns, jointly with Getty Oil, the Entertainment and Sports Programming Network (ESPN), which lost between ten and fifteen million dollars in 1981. Even so, to make sure that ESPN reaches its maximum audience, the network will pay cable operators ten cents per year per subscriber to carry its programming and, of course, its advertising.

ABC, in conjunction with Hearst, also offers ad-supported ARTS, carried on 1,500 systems to 7 million viewers. Focusing specifically on the performing and fine arts, ARTS offers such consummate fare as violinist Isaac Stern in concert, an exhibit from the collections of New York's Metropolitan Museum of Art, a documentary on the Leningrad Kirov Theater, and an exhibit of Russian paintings from the Los Angeles Museum of Art. CBS Cable was an all-cultural station, carried by 350 systems to 4.5 million subscribers. It offered programming in the performing arts, Broadway and off-Broadway plays, and art-related documentaries. Although CBS Cable expected to lose twenty million dollars in its first year of operation, it folded at the end of 1982 due to a shortfall in ad revenue. Nevertheless, all the major networks are getting involved with cultural programming whether through cable or pay television offerings. Cultural programming is attractive to the networks because, aside from its prestige value, audiences tend to be "quality," meaning in top income brackets, and thus prime targets for advertisers with expensive and sophisticated products.

Through its offspring Rainbow Programming Services, CBS offers, jointly with Cox Cable and Daniels and Associates, a pay service called Bravo to a limited audience of 50,000 people. Bravo, programming mainly foreign films and musical performances, has sent its crews out to tape performances at Carnegie Hall and the Indiana University Opera Theater; it also offers *Bravo* magazine, a *Sixty Minutes*–style cultural variety show. Bravo's fate is touch-and-go, for it needs 350,000 subscribers to break even. Its owners are optimistic, but there is plenty of competition in cultural programming. NBC ventured into culture via its parent company, RCA, in conjunction with Rockefeller Center. Its Entertainment Channel relied on the BBC's children's shows, comedies, dramas, and documentaries for about 40 percent of its fare; the rest is made-for-television specials, foreign films, and performing artists' specials. Subscribers had to pay a rather hefty twelve dollars a month, however, and the Entertainment Channel closed down operations in early 1983 after losing at least $34 million. Finally, not to be outdone, PBS has proposed its own cultural channel, to begin operations at a date when, it expects, a few of the other arts services will have fled the field.

Cable is potentially such a threat to the existing broadcast networks that all the television giants are committed to some type of alternate video programming. The networks have bypassed the FCC's stringent cable ownership regulations through complex webs of subsidiary, joint, and spin-off companies. ABC, however, deserves special mention for having figured out a way to eliminate its cable competitor altogether. Its Home View Network (HVN) is a nationwide pay service, offering movies, sports, concerts, and children's shows to subscribers for twenty dollars a month. HVN's signals are broadcast in scrambled form directly to subscribers' home videocassettes, which are equipped with special decoders. Although HVN's market is limited to the four million VCR owners, its programming is top of the line. Copyright problems are nil because subscribers can neither duplicate nor share their scrambled tapes with other VCR owners. As a result, movie companies make films available to HVN before offering them elsewhere.

Pay television, potentially the most lucrative of all the new video services, is as yet barely into the black. The big profit maker is Time's Home Box Office (HBO). After five tenuous years, HBO broke even in 1977, when it garnered 750,000 paying subscribers. HBO has now cornered nearly 60 percent of the United States pay television market with about 10 million subscribers and $75 million in profits on revenues of $315 million in 1982. HBO has bought exclusive rights to a substantial number of yet-to-be-released Columbia Pictures films; HBO also screens Broadway plays, variety shows, and some original programming. Recently HBO, jointly with CBS and Columbia Pictures, announced its intention to open a production house, the better to get programming exactly tailored to its own specifications. In 1983, Home Box Office had two of its own series on the air, five production commitments for series pilots, and twenty-six shows under development. Offering similar fare, Showtime, owned by Westinghouse and Viacom, a CBS spin-off company, expected to be operating in the black by 1985. Then, in the fall of 1983, Showtime merged with Warner Amex's The Movie Channel, which has about 2.4 million subscribers. The new team should provide stiff competition for HBO and its sister company, Cinemax. Like HBO and Cinemax, Showtime / The Movie Channel often outbids the commercial networks for the rights to Hollywood movies. Showtime also had six original series on the air in 1983 and ten more under development. In addition, some video services are experimenting with subscription television, sometimes called pay per view. A Sugar Ray Leonard fight and a Rolling Stones concert were made available to paying subscribers at an additional cost. The future will surely see more such events.

In Canada, pay television is just getting off the ground. According to CRTC regulations, 60 percent of the investment and program-acquisition budgets for pay television must be spent on Canadian programming. First Choice Communications, Ltd., for example, plans to spend $400 million over the first five years on Canadian programming for its pay service, including $120 million to acquire rights to all feature films made in Canada. Canadian programmers, always competing for their own country's audience against massively financed American productions, must be heartened by this news.

With more Canadian satellites in the sky, Canadian cable companies can expand into rural and remote northern areas. So far, the CRTC has approved 250 of the 1,400 cable applications from remote and rural communities. Canadian Satellite Communications, Inc. (Cancom), beams signals to earth stations; officials predict that by 1984 its earth stations will make programming available to even the tiniest, most remote communities. These new earth stations will create jobs for northern residents, many of them native peoples, installing and maintaining equipment. Cancom's radio programming in the Indian and Inuit languages will create programming jobs; television programming in native languages should soon follow. Cancom eventually hopes to market its Canadian programming, in French and English, to American cable companies serving areas with large expatriate Canadian populations, such as Florida and California, or areas with many Americans of French extraction, such as Louisiana and Maine. Such plans cannot but be a boon to the many now working in media-related jobs in Canada: 30,000 in broadcasting and cable; 30,000 in the independent film and television industries; 10,000 self-employed on film and video productions as performers, writers, technicians, and the like.

Programming activity in the United States and Canada is already nothing short of tremendous. Bearing in mind that as yet only 30 percent of America's eighty million television households have basic cable and that technology promises far more video capacity in the very near future, one begins to see why programming is such exceedingly fertile territory for the job seeker. Simple arithmetic points out the truly staggering potential of programming. If we assume, conservatively, that sixty channels operate twenty hours daily across the nation, then 1,200 program options will be necessary *every single day*. Subscribers to the new video services are not paying to see reruns. If we multiply that 1,200 by a conservative 200 days a year (or even 100), we see that 240,000 (or 120,000) hours of *original* programming must fill that time. In 1982, forty-two satellite and thirty-five cable programmers supplied cable operators, with another twenty already scheduled to begin distribution by 1984. Somebody will be working all that video equipment; somebody will be making all those programs; somebody will be performing in front of all those television cameras. Will it be you?

Traditionally the broadcast networks supplied almost all the jobs in television. Today it is next to impossible to find a job there. Dick Plante, Personnel Director at CBS headquarters in New York, estimates that CBS has only 2,000–3,000 openings a year, and more than half of those positions are filled internally. He regularly receives ten times that number of applications. CBS generally hires only those with the top grades from the top schools. Many have MBAs and a handful (about forty) have bachelor's degrees in computer science. Due to the company's laudable policy of giving priority to internal applicants, many secretaries have gone on to bigger and better things. But word to that effect went out long ago: as a result, CBS has an amazingly large number of male secretaries.

The new video technologies have opened up many alternate paths to a career in television. Programming departments in local broadcast and cable stations require control-room technicians to operate cameras and videotape recording equipment, film editors, sound technicians, production supervisors, and graphic artists, to name a few. Even if you don't have a degree in video communications or television broadcasting from a community college or a university, though it would help, don't despair of getting into the industry. At Ted Turner's CNN in Atlanta, unskilled applicants, usually university graduates who show promise, are hired as "video journalists." The position involves three months at the minimum wage while training in all aspects of television production. After that, according to Turner Broadcasting Systems' Personnel Director William Shaw, the trainee moves elsewhere or becomes a production assistant; within a year, that person might become a director or producer. Turner Broadcasting hires five to ten applicants a week for its various services.

By order of the FCC, every cable franchiser must reserve public access channels for local educational institutions, public service groups, municipal government, or just about anyone else who wants to use them. Thus every cable system provides another training ground for the industry. For less than it costs to take tennis lessons, the cable system will train you in a brief course to operate video equipment and create your own program, which you and every other subscriber in your system can then view on your home screens. Indeed, through these public access channels you could develop a hobby or get involved part time in what might eventually turn into a paying position and a new career. Larry Rawson, a broker at Paine, Webber in Boston, who was also a runner, watched a broadcast of his city's famous marathon. The commentator's misinformation so irked him that he drove down to the press box and saw that it was corrected. Five years later, he was himself the commentator for the first televised coverage of the marathon on public television channel WGBH. He sent a videotape of his first performance to the nascent ESPN, and now he's their track and field commentator.

It happens all the time. With a little hard work, a little creativity, and a little *chutzpah,* it could happen for you, too. Viewers love to watch public access programs, perhaps because of their spontaneity and unpredictability. New York City offers often hilarious and exotic entertainment over its two public access channels: *The Womb of Aquarius, Jimmy Dee's Peeping Tom Show, Telepsychic, Van Gogh Video, Innertube,* and *Artifact* are a few titles. In East Lansing, Michigan, WELM attracts 5 percent of its 20,000 subscribers to its public access channel, although it competes against twenty-four other stations. An independent survey taken in Bloomington, Indiana, determined that 50 percent of the local cable system's 9,000 subscribers watched public access Channel 3, especially to see *Kids Alive!*—produced by eight- to fifteen-year-olds. One enterprising fourteen-year-old New Yorker produces *Mark Schlictman's Kids News.* Many high school students now televise their own team sports over local cable.

Another little-known way to learn the business is at a low-power station. In 1982, the FCC agreed to license 3,000–4,000 such stations across the United States, at a cost of about $50,000 each. Low-power transmission, a technology as old as cable, can amplify and rebroadcast faint signals within a ten–fifteen-mile range. Until 1980, FCC rules forbade the stations from originating programming, but wanting to make access to public airwaves more democratic, the FCC lifted the restriction. By 1983, over 7,000 applications for low-power stations had reached the FCC which had given the go-ahead to one hundred, granting preference to groups traditionally underrepresented in the television industry, such as minorities and women. Michael Ice, a black applicant, wants to offer programming to Chicago's black, Mexican, Asian, and Polish populations. Special interest groups, such as religious organizations and independent political parties, are equally excited about the possibilities. Since many individual applicants hope to network with other low-power stations, some groups have made multiple applications. Three former FCC attorneys have applied for enough licenses to launch a nationwide black network. Eventually various low-power stations could be linked by satellite. Broadcast networks are in on the action, too. ABC and NBC have applied for licenses. The Arizona-based Neighborhood TV Company, nearly 50 percent owned by Sears, Roebuck and Co., has applied for 141 licenses and is planning a nationwide satellite-linked network offering country and western programming.

FCC spokeswoman Rosemary Kimball expects that each station will employ about five people to operate and maintain the equipment. Your nearby low-power station could be the place where you get your start in television. Jim Boler, a retired broadcasting executive, is the owner of an interim-licensed low-power station in Bemidji, Minnesota. Through a door-to-door survey, he found that people were interested in local newscasts, high school sports, and country and western entertainment. So that's what he broadcasts. Most likely other small communities will want similar fare. Here's your opportunity to become a reporter, a weather forecaster, or a television entertainer.

Non-broadcast video production is yet another entryway into the television industry. Non-broadcast video has an added advantage: though hardly known to the outside world, it is itself a multimillion-dollar industry. Leading *haute couture* houses, for example, videotape fashion shows and send them to retail buyers around the world. On short airline flights, small features covering, say, the story behind the Rose Bowl are shown. Most universities and hospitals have videotape production units, which serve many purposes, such as recording an unusual operation, a lecture by a famous visitor, or even demonstrating the institution's layout to new staff. Corporations and businesses account for much of the non-broadcast industry's spectacular 40-percent-a-year growth rate. It's much cheaper to put the company's president on videotape than on an airplane. Videotape productions also orient new employees, explain safety procedures, and motivate managers. In 1972, private industry produced more videotaped shows than ABC, NBC, and CBS combined. Since then, the business has mushroomed.

Michael Gowling, who owns his own video production house in Toronto, thinks of himself as a professional communicator. His concern is facilitating communication within an organization rather than slick programming. He uses videotape because it is cheaper and faster than film, plus tape is reusable. Since he works mainly for large corporations, he also likes video because it has greater psychological impact with all the intimacy of the familiar television screen. But he would "go into alpha waves tomorrow, if that's the best way to communicate." His latest job was a seven-minute, $60,000 computer animation project. As a one-person production company, he wrote, produced, directed, edited, and did the camera work. Sometimes he uses outside help, of course. Like many entrepreneurs, Michael values the autonomy of being his own boss; he finds he has room to grow and experiment.

Although he left school at sixteen himself and learned all he knows about video production through practical experience, Michael teaches a course on the art of corporate communication at nearby Ryerson Polytechnical Institute. Few of his students are aware that non-broadcast is every bit as professional as broadcast video: Michael uses the same sophisticated video production facilities as the Canadian national broadcast network, the CBC. As vice president of the Toronto chapter of the worldwide International Television Association, he naturally advises joining the group to get started in the industry. Members receive monthly newsletters full of information about what's happening in the business. Another way to enter the industry is to find a job in a large corporation; large corporations usually have in-house video production departments. A true entrepreneur at heart, Michael Gowling isn't too keen on that alternative; he finds most corporate production types too technically oriented, with "video tunnel vision." Michael has ideas. Here's one he'll give you. VCRs have as-yet-unused narrowcasting potential. Why not make commercial self-help tapes the buyer can consult over and over again? How about car repair or plumbing? Aerobic dancing? Yoga? This industry has room for all kinds of ideas.

The possibilities generated by the new video technologies are so vast that no one should feel left out. There are opportunities for everyone. In Minneapolis, two cerebral palsy victims wrote and produced a documentary about the daily lives of quadriplegics for public access. In Bloomington, Ron Markman took public access viewers on a tour of his personal fantasy land, Murfa. *Black Entertainment* is offered to 9 million subscribers through 850 cable systems. The Christian Broadcast Network preaches the Gospel through television. Warner-Amex's Nickelodeon services provide children with noncommercial, nonviolent, non-sexist, and non-racist information, education, and entertainment. The Modern Satellite Network has women's programming; Cable Health Network appeals to the hypochondriac in all of us.

For the first time, North Americans have a window to other cultures and language groups through foreign and foreign-language programming. The SIN Television Network is a nationwide service all in Spanish, with some programs received directly from Spain or Latin America by satellite. For unilingual Americans, it could be a living language lab; for Hispanics, it's a way to participate in American life on an equal footing. Toronto's Channel 47, the multicultural station, offers a smorgasbord of programming eighteen hours a day: *Islamic Horizons, German Carousel, Black World, Filipinesca, Chinese Magazine, Portugal Today, Italianissimo A.M., Iron Samurai, Middle East Review,* and foreign movies serve the city's many ethnic communities. In the United States, Radio Telefis Eirann, the Irish national broadcast service, has become a cable programmer. Distributed by cassette to cable systems, *Ireland's Eye* airs two hours a week over the Satellite Program Network (SPN). SPN also has *Telefrance-USA, Studio 1* from Italy, and *Vision of Asia-USA* fom India. Montréal's French-speaking population is so large that the city's Cablevision Nationale' Ltee. can air on Channel 99 the full program schedule of a regular broadcast network brought directly from Paris by satellite.

Naturally many people will find steady jobs with particular programs, but a great deal of steady free-lance work will also be available in the industry. Every television show, no matter how it is transmitted, generally requires several of the following professionals:

- producers
- directors
- script editors
- writers
- researchers.

Shows also need workers with more specific skills, such as:

- set designers
- costume designers
- makeup artists
- hairdressers
- light and sound technicians
- film editors
- special effects artists
- graphic artists.

One person might fill three or four of these positions for a small-budget show. Producers, for example, not only conceive the program idea but are also frequently responsible for finding a buyer. Producers also arrange financing, hire personnel, find locations, and may even research, write, and edit the script.

Dawn Carpenter (not her real name), at twenty-six, is the producer of a popular news show on a network-owned local affiliate in a large American city. Although she has a BA in English literature, she got her start at twenty while working as a "coffee cup girl" at the affiliate. She progressed from writing copy in her spare time to writing scripts to occasionally performing on the air and then to producing special-event programs. According to Dawn, her station hires people "off the street," provided they have good looks and the right personality. In her experience, broadcast television is one industry which, although run by men, does offer opportunity to women. Nor does one need a lot of education to get started. It is mostly luck at first, then hard work and more hard work. It is a young person's business.

Jancey Ball is a free-lance producer who came into the industry through the technical end, a most unusual entryway for a woman. At New York University, she studied history but worked part time on the lighting and sound for local rock and roll productions. She gained production skills at the Berkeley Film Institute, which offered its students hands-on experience. In California, she offered her services free of charge to a group making a small educational film explaining the Pledge of Allegiance to children. The experience was invaluable. To make money, she worked in a film laboratory, running the sound department. One memorable summer, she and two thousand others staffed the Renaissance Fair, a re-creation of an English country village north of San Francisco. With its Living History Center, five stages, and 20,000 daily visitors, it was an organizer's nightmare. She ended up with the title of assistant to the producer, and she did everything—and learned everything. Since returning to New York three years ago, she has produced educational and industrial films and worked for the CBS subsidiary Viacom, where she was the only woman above the secretary rank. The skills Jancey needed for all her production jobs were basically the same: an ability to deal with people, unions, insurance companies, and truck drivers. It helps to know the machinery, but if you're good at judging people, you can always hire the best technicians. Being a free-lancer is hard and very insecure, and you need to have many contacts in the industry, but you can do what you like.

Producing a program is, of course, only part of the process. Someone has to market the finished product. Often this someone is a television packager who coordinates all aspects of a show, from inception to distribution. The packager first secures an option on the proposed property and then finds a producer, a director, writers, and actors. Finally, through business contacts, the packager ensures the widest video distribution. For his or her efforts, the packager gets a percentage of the take. With the multiplicity of channels now available, a new job title, video agent, has appeared. Familiar with the program marketing and the distribution networks, the video agent finds buyers for an independent producer's ideas. The video agent often doubles as a packager. Syndicator is another position in programming. The FCC has ruled that broadcast networks producing original programs cannot resell or syndicate old shows. Though the networks are campaigning vigorously to change this rule, at the moment either independents or network spin-off companies like Viacom perform this function. The voracious public appetite for cable programming is making syndication ever more important. The production or programming department of a cable or pay television station also includes personnel responsible for evaluating program ideas and purchasing them. Other staffers are program coordinators, who line up the dates and times when the shows will air.

Performers have the most glamorous jobs of all. Intensified programming activity is sure to benefit the nation's starving artists. Singers, actors, dancers, comics, impersonators, voice specialists, musicians—entertainers of every stripe are sure to benefit from the explosion in television programming. As a result, according to Harry Smith, vice president for technology at CBS, many supporting entrepreneurial endeavors will prosper. The fashion and beauty business, for example, is vital to creative performers. Acting, modeling, and elocution courses will be important not only for professional performers but also for the many ordinary people who will be appearing on television. Consultants who help people look and act their best in front of the video camera should become more numerous. Agents, photographers, promoters—all these and many more will experience an increase in business.

Though many industries will benefit from television's expanded programming, the one with the most to gain is the movie industry. Once upon a time, television was considered the upstart rival of the motion picture industry. The fact is that television has become a major market for films. Cable and pay television are pushing that collaboration into high gear. By the early 1990's, it is highly likely that television's expanded programming will be responsible for the movie industry's main source of revenue. The signals are already evident: Twentieth Century-Fox and the ill-fated CBS Cable were partners; Paramount, Universal, and the USA network have joined forces. In the next few years, movie companies must get production, marketing, and distribution costs under control. Pictures now costing ten million dollars to produce and another four to five million dollars to market must come down in price. With revenues guaranteed by a cable or pay television service, films can be made on smaller budgets for smaller audiences. Film makers can explore themes which are commercially risky for the mass market. By the 1990's, two or three times as many films will be made. That will be the Golden Age for writers, directors, actors, actresses, and all other talented performers.

Surprisingly the recording industry also stands to gain from expanded television programming. Warner-Amex has already launched Music Television (MTV), twenty-four hours a day of rock and roll television, currently carried by 300 cable companies to about 3 million viewers. MTV shows "pop clips," technically sophisticated, fantasized song impressions starring the musicians. According to media consultant Bob Klein of Los Angeles, pop clips are the future of the music industry and a fascinating little art form in themselves. Combined with impending technologies like stereo television, televised music may soon be video's next big success story, generating jobs for directors, conceptualizers, producers, and performers who will one day develop it into a business larger than today's recording industry. Pop clips point the way for pop music to become a more visual art. As a result of rapid technological developments in home entertainment systems, film producers and record producers will inevitably merge their resources, trying to satisfy increased market demand for audio-visual products. The days of rock and roll groups plunking guitars while jumping around a stage are numbered; more performance-oriented musicians are already taking their place. Jazz, opera, country and western, and other kinds of music should follow the lead of pop. Local music stations on cable will be standard by the 1990's. Musicians will no longer go begging for jobs.

As in other high-tech industries, the software side of television has great potential. The resulting "soft" jobs will be numerous and open to people from diverse backgrounds. Already, advanced television technology has liberated the home viewer from thralldom to the networks and vastly expanded the programming sector. The new media's ravenous appetite for programming has also, not incidentally, stimulated intense activity in supporting industries. By the 1990's, the action in programming will be nothing short of frenetic. Those who are aware of this industry's potential today are sure to create even more surprises for us all in the future.

CHAPTER SIX

INTERACTIVE TELEVISION

In the past the communications industry was a clearly defined enterprise which included the telephone, the telegraph, and Telex. The information industry—data processing—grew up as a result of the computer revolution of the past thirty years and is now a well-established fact of modern life. What is not well understood, however, is a recent change in these industries which *has already occurred.* Communications and information technology have merged. There is no longer any real distinction between a computer terminal used for communication and one used as an input-output device for data processing. The markets of the two industries are becoming one, served by both telephone and computer companies. The area of most obvious overlap is communications-based information systems, such as interactive television or videotex.

The revolution caused by the introduction of the electronic media, like that caused by the introduction of the computer, occurred only recently, but its effects are here to stay. The influence of television is everywhere in our lives. Digital technology is making changes in the television industry as elsewhere. Because of their common use of digital technology, the telecommunications sector, the computer sector, and video technologies are blending into one massive industry. This industry has been called by various names—in the United States, compunications, an unwieldy term; in France, *télématique;* in Canada, telematics, a more pronounceable word. Whatever it is called, this mega-industry will have enormous impact by the 1990's.

VIDEOTEX

At the vanguard of the convergence of the communications, information, and video industries is a most provocative new technology called videotex. Videotex, a kind of two-way interactive television with a wide range of potential applications, can make a feasible reality out of what are now merely contemporary catchwords such as "the wired city," "the cashless society," and "soft cash." Electronic mailboxes, teleshopping and telebanking, electronic libraries, access to distant data banks, and many kinds of graphic and photographic reproduction can all become instant realities through the wonders of videotex.

Invented about ten years ago, videotex technology has advanced greatly during this initial stage of development and increasing sophistication; the so-called fourth generation went into production in 1981. As yet, videotex has become a viable business venture accessible to the general public in only one country—the United Kingdom. Three countries, the United Kingdom, Canada, and France, have invested heavily in competing systems, called Prestel, Telidon, and Antiope respectively. The race is on to see which country will win what share of the market.

Technically videotex could be marketed to the general public in all three systems immediately. The main reason for the delay in putting this technology into the hands of the consumer is not technical. With appropriate modifications, videotex is capable of so many applications that no one wants to invest in a particular system before consumer acceptance and need can be demonstrated. Thus field trials and marketing studies are now being completed in many countries, including the United States and Canada, before production decisions are made. Throughout the eighties this new market is expected to grow by 15 to 20 percent per year. Thousands of videotex terminals were in operation by 1982; hundreds of thousands will be in operation by 1990. By the end of the decade videotex will be as familiar and as ubiquitous as the television set is today. The result will be changes for all of us.

WHAT IS VIDEOTEX?

Videotex is almost completely unknown in North America. As yet most of us have never seen it in action. Although all contemporary systems are similar, Telidon, the Canadian version, is currently the most sophisticated and versatile of all. As it seems likely that some form of Telidon technology will be the most widespread in North America, it is Telidon technology which is specifically described unless otherwise indicated.

Videotex technology allows specially adapted television sets (or display monitors) to access a computer data bank and call up textual or graphic material onto the screen. The system is therefore called interactive. The central computer can also act as a switch to relay requests and retrieve information from third-party data bases in a remote computer. In addition, the system allows direct terminal-to-terminal contact with another user, and it can serve as a minicomputer for home or office use.

The main components of the system are a television set modified with a decoder (or modem); a communication carrier, which can be telephone lines, television broadcast signals, coaxial cable, fiber optics, or satellite; and a computer. Attached to the display monitor is a special interface (the decoder) which receives signals from the computer and converts them into text and images on the screen. For home use, a key pad, or for business use, a keyboard (like a typewriter) can be wired to the device or operated by remote control.

Schematically the simplified system looks like this:

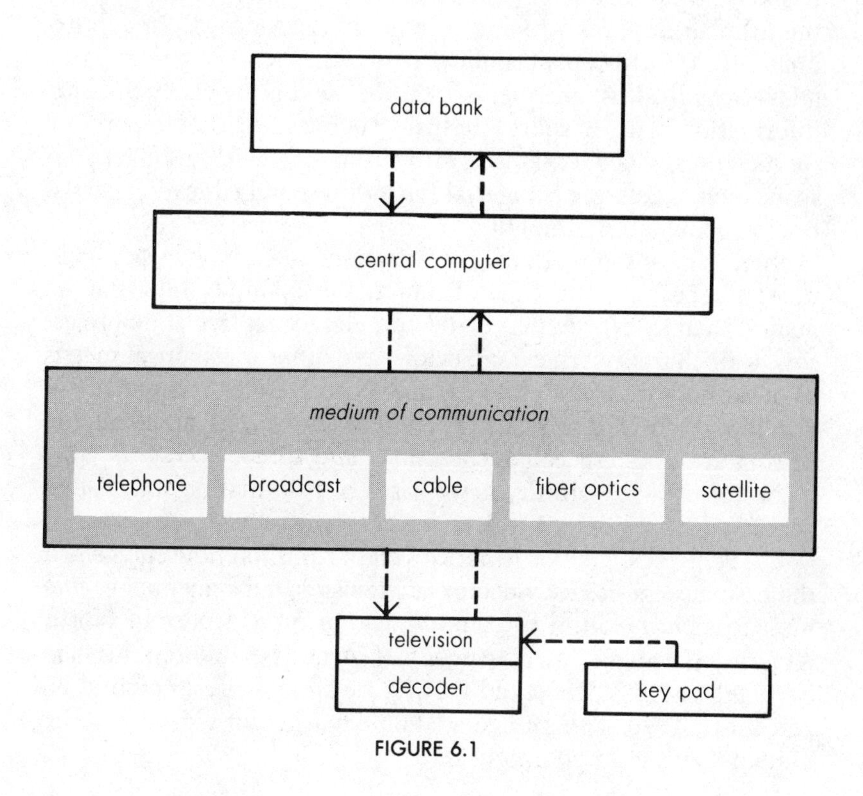

FIGURE 6.1

The terminal could also be a home computer, either a desk or a hand-held version, equipped with the appropriate peripherals. Depending upon the needs of the user, the basic schema could be modified in any of a variety of ways. For example, a digital facsimile service, a hard-copy printer, or an information provider terminal could be added. Other adaptations will be discussed below.

Teletext is the name given to the broadcast mode of videotex. This one-way system permits the user to receive text and graphics on an ordinary television set equipped with a decoder. The signal is broadcast in digital form over the air or via cable in the unused section of the regular television signal, called the vertical blanking interval, or on a full television channel. Closed captioning for the deaf is one familiar example of teletext. In a hybrid or noninteractive system, the user requests information from the television station and then must wait until it is scheduled for broadcast. Response time is generally slower than in a videotex system, and the number of pages which can be broadcast ranges from approximately 100 to 15,000, depending upon whether a partial or a full television broadcast channel is used. Teletext is generally used for information with a short life-span, such as weather reports or sports scores; videotex is used for information with a longer life-span, such as reference material for business and education, statistics, or home entertainment.

Increasingly sophisticated modifications to basic videotex technology appear all the time. Telidon, for example, now has an audio-visual system which could completely replace slide projectors, wall charts, overhead projectors, and other mechanical means of presenting information to groups. Plans are also being made to develop audio capability in the system, as well as photographic editing routines, speech recognition, and direct access to large blocks of data containing many pages of information for off-line perusal.

In 1983, 98 percent of America's eighty million households had the basic necessities for videotex, a television and a telephone. The most feasible medium of communication for videotex in North America is cable. Thirty percent of American households had cable television in 1980 and that figure should rise to around 60 percent by 1990. The immediate home market for videotex in the United States is an enormous one.

USES OF VIDEOTEX

Check the daily weather report. Buy a pair of running shoes for Saturday's marathon. Transfer funds from one bank account to another. Wish a friend "Happy Birthday!" Plan and pay for a holiday trip to Tahiti. Look for a job. Read a story to a child. Discover the best fertilizer for the carrots in the garden. Find out your horoscope of the day. Without leaving home, you can do all this and more on your videotex terminal.

Sound impossible? Theoretically, current videotex technology can supply all the above information and services. Of course, the information to be accessed and the function capability must first be put into or provided to the system before it can be taken out by the user. At present no one system has been prepared to supply everything mentioned above. One could have been, but it has not been, because of marketing, not technological, constraints. The British system, the only one in commercial use, offers a wide range of viewer services with the greatest number of "pages" for display. To list everything obtainable would be tedious, so some examples will have to suffice. On this system it is possible for a fee, generally a few pence a page, to access a wide range of business information—such as up-to-the-minute stocks, bonds, and commodities prices direct from the stock exchange. Job vacancies in the information industry are listed, as are real estate offerings. An entertainment and theater guide for the London area is available. A publisher presents a different children's story every day complete with accompanying graphics. Selected businesses are profiled. The subscriber can choose from a wide selection of video games. A teleshopping service is available, and payment is made by punching in a VISA or AMEX card number. An electronic newspaper presents the news of the day. These are only a few examples of the information and services made available by this pioneering commercial venture—which in turn are only a small percentage of those the system could potentially offer to the consumer if it were not for financial limitations. Although it has not yet reached the break-even point, the British system has proved consumer acceptance and enthusiasm beyond a doubt, and it is expected to show a profit within a short time.

All extant videotex systems employ what is known as "user-friendly" machines. This means that the system is designed to make it easy for anyone, especially technophobics, to use. No knowledge of computers is required. Typing skills are unnecessary. Anyone who can read, follow a simple index logic, and push buttons can use the system. And as difficulties are revealed through widespread testing, even routing methods used to dialogue with the index are constantly being streamlined.

Another example of the actual application of videotex is supplied by the French Antiope system. Now in the advanced field trial stage, the French have provided their system with a great number of service functions. Antiope is operated by the state-owned telephone company under the auspices of the French government, which has given careful thought to the future development of videotex in that country. This system offers the electronic mailbox: the user can either send preselected greetings, such as "Bon voyage" or "Get well," with accompanying graphics or leave a message. The subscriber's incoming messages appear on the display monitor at the push of a button. Personal banking can be done on this system, such as transferral of funds from one account to another with an instant readout of funds remaining in all accounts. The system also functions as a minicomputer to do home finances. Eventually the French telephone company plans to deliver all directory information through videotex and cease printing books. By the end of the decade, every telephone subscriber in France will have a videotex terminal installed gratis for this purpose. The current field trials offer directory information for selected areas only.

Both the French and British systems, as well as the Canadian one, offer teleshopping, but the French have added a new twist, the "smart card." The smart card looks like an ordinary credit card, but it is equipped with a tiny microprocessor with a 4K memory, and more memory capacity is planned for the immediate future. The microchip provides an electronic key to a confidential coded data file which is the card's memory. When the card is introduced into a payment machine and a personal code entered, the memory identifies the bearer, indicates the money value, or banking power, of the card, and automatically dates and records the information within the circuit. An average smart card transaction takes about thirty seconds. Thus the card serves as a personal identity card and provides almost absolute security, unlike the ordinary credit card, which must be laboriously verified for each transaction. If the smart card is stolen, for example, and the thief does not provide the correct secret identity numbers when trying to use it, the card self-destructs.

There are many potential applications for the smart card—as a high-security access card, a health card, or a social security card, for example—but two would be particularly far-reaching in their consequences. The card could be used in making a transaction with a point of sales (POS) machine located, say, in a supermarket—in other words, a "soft cash" register. This application would, of course, revolutionize the retail system as we know it today and truly introduce "the cashless society." In addition, the card could be inserted into a simple slot attached to the home videotex terminal to make instant telepayment for goods purchased through teleshopping. Several hundred of the videotex terminals in the current French field trials are equipped with smart card readers. According to Roy D. Bright, managing director of Intelmatique, the international marketing arm of the French national videotex program, use of the smart card provides "an altogether higher level of security for personal financial transactions over the videotex system. It virtually eliminates the possibility of impersonation or fraud, making it perfectly feasible to access personal bank records and other sensitive data."

The smart card is generating enormous interest beyond the boundaries of France, and the reasons are not hard to fathom. One of the hallmarks of the age of "the information explosion" has been the corresponding ballooning of paper work. Millions of checks and credit card transactions are processed through banks and private financial institutions every day, creating enormous expense in overhead. The people who keep track of these accounts must be paid. The paper must be stored or the information recorded and the paper disposed of somewhere. It has been estimated that every check written costs the originating institution one dollar in global overhead. Using the smart card, banks and eventually the consumer could save millions of dollars while gaining a higher level of service. If the consumer accepts the "soft cash" concept, as seems likely, our daily business life will soon look quite different from the way it does today. And videotex terminals will most assuredly incorporate a new modification as part of the basic system.

It must be emphasized that videotex has almost unlimited potential applications. The services listed above represent only a part of what videotex technology is now capable of supplying, not to mention what might be in store for us in the future. So far, videotex technology has unfolded with dizzying and ever-increasing speed. However, due to marketing considerations, whether, when, and how videotex becomes an economic reality in North America are still open questions.

THE FUTURE OF VIDEOTEX

There are many reasons for the uncertainty over implementing videotex in North America. Part of the uncertainty arises from the problem of public acceptance. Will consumers welcome videotex as they did television? Another problem lies in the confusing array of incompatible systems. No one wants to sink huge amounts of money into the wrong hardware.

A major part of the uncertainty, however, stems from the host of complex social and policy issues which this revolutionary technology will raise. Public use of computer-based information will profoundly alter our society, our industries, and our professions in ways we can only guess at. Some of the immediately obvious questions which must be answered have to do with

- ownership of the system
- source of the information
- role of the telephone system
- role of the cable television systems
- role of television, both public and private
- privacy and security
- cost and methods of payment
- impact on the banking system and the post office
- social considerations: will it lead to more alienation and isolation?
- role of government and regulatory and other public agencies
- impact on the economy: short- and long-term labor dislocation or displacement; implications of transborder data flows on the balance of payments

- impact on the current communications industry, e.g., advertising revenue for newspapers, magazines, television, etc.
- impact on the professions
- question of copyright and ownership of information.

These questions have not yet been answered and the technology is already upon us. Time is running out for field trials, investigating committees, and public commissions. The complexity of the issues helps to explain why a go-slow approach is being tried here, but such a potentially giant industry cannot be halted for long. The profits to be made are just too big. The new jobs to be created are just too many. North America cannot afford to lose this opportunity while it dithers over implications and ramifications. If videotex is not introduced in a rational and controlled manner, the lure of enormous profits will force it in in a haphazard and chaotic manner.

Other countries, notably France and the United Kingdom, have already addressed and solved many of the thornier issues, each in its own unique way. The situation in France provides a clear example of how videotex can be introduced while at the same time maximizing the benefit to the economy as a whole. In France the government has already made firm decisions. In ten years the French government has rushed to modernize its telephone system, which in 1970 was among the most inefficient and antiquated in Europe. The French chose to systematically develop the most sophisticated and advanced techniques. Along the way, the government realized that France had a good chance of becoming a world leader in the new field of telematics. Through careful planning, it has encouraged the growth of the computer and electronics industries. Centralization of decision is most important in France, as is government support for fledgling industries. According to French plans, videotex will have created a huge new market for domestic industry by the end of the decade.

French videotex will be in the testing phase for the next couple of years with the major impetus planned for after 1985. By that time, the national telephone network will be largely digitized, and optical fibers will have replaced many of the cable links. This will enable subscribers to take full advantage of a wide range of telematic products.

French industry is expected to produce mass-market equipment such as videotex, teletext, telecopiers, and POS terminals. The enormity of the growth in this one medium-sized country can be seen in these figures:

	today	*1985*	*1990*
telecopiers	5,000	300,000	800,000
videotex	270,000	8–10 million	25–30 million
POS	nil	25,000	?

To facilitate the growth of its telematics industry, the French government expects to spend twenty-five billion dollars during the 1981–1985 period. This is an enormous figure for one medium-sized country to invest. When one considers that spending by private industry will at least match and possibly double or triple that figure, it is obvious we are talking about very big money indeed. Clearly, French financial planners believe that the potential profits and job creation opportunities generated by this industry justify such an enormous outlay, though undoubtedly the country's current economic tribulations will force some modification in the government's projected program.

France is a relatively small, though economically important, country. The virtual certainty of the widespread use of videotex there and in the U.K. makes it all the more inevitable that the United States and Canada will also adopt the technology. The United States is, as always, the most important market in the world. Acceptance of videotex in the United States means billions and billions of dollars. Initially business use of videotex will far outstrip any other. But every television owner is a potential customer for videotex. In ten years those who have not even heard of videotex today will consider it an absolute necessity for their homes. Spending on services and equipment will be staggering. A mighty new industry will be created, and it will generate thousands of exciting new jobs.

THE VIDEOTEX INDUSTRY: JOBS

As videotex technology is implemented in the 1980's and early 1990's, the various sectors of the infant industry must display a good deal of overlap. In the beginning, as the first-time demand for hardware is satisfied, the technical and *manufacturing sector* will have a preponderant role to play. At the same time the *service sector* must develop the information which is the whole purpose of the industry. The *support sector,* which includes administration, sales, and marketing, must be sure that users have the equipment and access to information appropriate to their individual needs. None of the sectors could exist without the others. All must share in the prodigious effort needed to educate the public in a totally new technology.

MANUFACTURING SECTOR

Videotex is now entering the preliminary stage of commercial mass-market manufacture. Manufacturing the equipment to meet first-time demand will be the work of the 1980's. Thus most job creation in the manufacturing sector will be over by the mid-1990's. Videotex is expected to follow the path blazed by television. The first phase of manufacture produces the equipment, which is distributed as widely as possible. Then the role of the manufacturing sector changes. Now the emphasis is on gradually introducing equipment which is more sophisticated and/or specialized, the manufacture of software, and technical maintenance. By the mid-1990s, the manufacturing sector will have entered the last phase.

Although the videotex manufacturing sector will have diminished in importance for job seekers in the 1990's, it will nevertheless continue to give impetus to computer-related industries. It is worthwhile to mention, however, that demand for certain specialists will remain strong in videotex manufacture as well as other computer-related industries. Among them are:

- information scientists of all types
- telecommunications engineers
- hardware/software engineers
- data processing experts
- semiconductor technologists.

Further, systems analysts, particularly those specializing in minicomputers and the integration of word processors into computer networks, are of special use in videotex.

The plans of the French government to put twenty to thirty million terminals into use by the end of this decade will bring into prominence yet another new field of technology. For it will be impossible to accomplish this task without increased automation: enter the expert in robotics, the use of machines (robots) to manufacture machines. Robotics, of course, has applications in many other types of manufacture.

In addition, several other new fields will play critical roles in the development and refinement of videotex. Computer-aided design and manufacture is now in an embryonic state on this continent, but is widespread in Japan. As videotex enters its third phase of manufacture in the 1990's, the production of specialized equipment will be essential to the continuation of the industry. In fact, users will demand design customized to personal specifications. The only way that this can be done is through computer-aided design—which will require the services of industrial designers skilled in computer applications.

In the next decade, as interactive computer-based information systems spread into every home, computerphobics will have an exceptionally difficult time. And the handicaps of the unskillful or fearful user will have to be dealt with. This is the task of ergonometrics. Ergonometrics is the study of the man/machine interface. Ergonometricians are those who make user-friendly machines ever more friendly. They will be extremely important if the "wired city" and the "cashless society" are ever to become realities.

SERVICE SECTOR

Although manufacturing videotex hardware will be useful in stimulating the micro-electronics industry in many countries, the primary purpose of the technology is to provide information. The service sector of the industry will be responsible for creating this information. This sector thus has the greatest long-range potential for the job market. It also offers the most intriguing variety of possibilities to the job seeker.

Content in videotex is anything, including services, which can be stored on a computer and retrieved by means of a videotex key pad. Organizations or individuals who provide content for videotex systems are called information providers, or IPs. Those who deliver specific information to specific markets are called umbrella IPs, or information brokers. Those who create the information for this technology will have to take into account the fact that the method of presentation of videotex content is different from that of any other medium.

Exactly how IPs on this continent are to be organized is impossible to say. Much depends upon how the new industry is structured. The structure of the industry in Britain and France, where videotex is most advanced, is strikingly different in each country. In the U.K., where it is already a commercial venture, it is handled as private enterprise. In France, where its development is more embryonic, it is essentially government controlled. For the United States with its long free-market history, the U.K. model seems the more appropriate. So it is to the English example that we will turn here for information about jobs in the service sector.

In the U.K. the videotex system, called viewdata there, is operated by the British Post Office, which controls the telephone system, the carrier for viewdata. The subscriber pays a telephone bill each month, to which is added a charge, billed at the local call rate, for the hours of viewdata service used plus a small fee for billed pages. A page is the amount of videotex information that can appear on the screen at one time. Many pages are free. It is recognized that this is a clumsy method. Cable television, which has a much higher transmission capacity than the telephone line, is widespread in North America and is the most favored candidate for carrier.

In Britain there are about ten top information brokers. They are generally divisions of already established communications companies or publishing houses. The function of these umbrella brokers is manifold. They not only create pages of information themselves, like any IP, but also

- market the information for others
- provide consultancy services to client IPs
- train IPs
- educate through in-house workshops and conferences.

Education is an important part of the service offered by an umbrella IP. These organizations show client IPs how to structure and set up information, how to use equipment, and which equipment best fits their needs; they also engage in publishing hard copy, printed texts, and "micro publishing"—of microfilm and microfiche—as well as various combinations thereof.

About two thirds of all IPs in Britain currently use the services of one of the broker or specialist companies. With a few large companies to provide technical and marketing expertise, almost anyone can buy or rent the easy-to-use equipment and become an IP. In fact, on the British system a husband and wife team has been offering a London theater guide for over a year. The duration of their involvement suggests that they find the business profitable. A Mom and Pop information boutique may just be *the* small business of the nineties.

What exactly are the jobs available, and what kind of a background do you need to get one? According to Paul McFarland, manager of the viewdata division for Mills and Allen, one of the top information brokers, he is looking for people who are "young and smart. It's important to be flexible and open since we don't know where the industry is going. The technology is totally open." At twenty-six, McFarland is one of the "old" men in the business. After completing a degree in medieval history, he spent a few years as a sportswriter for a small newspaper and then answered an ad. . . . Progress is rapid for the creative and energetic in this industry. There are no experts. Most of the young people on McFarland's staff have arts degrees or a background in journalism. No one needs management training or a computer background. All training is done on the job. Aside from enthusiasm and flexibility, the single most important qualification is literacy.

The people in the service sector must perform a variety of tasks. As individual IPs, they must

- plan programs and originate material
- design routing systems
- input/output data
- edit and update data
- design graphics and page layouts
- manage and supervise
- train others.

Usually, of course, no one person is expected to perform all these tasks. Rather, groups of individuals work together to produce the final product. As in the other electronic media—television and radio—teamwork is essential.

It is well known that every communication medium has its own method of presenting information. Writing for television is not the same as writing for a magazine or for radio. Videotex is no exception. Although the job title may be a familiar one, such as writer, editor, production manager, or script supervisor, the jobholder will use his particular skills in a manner appropriate to videotex. Because videotex has graphic capability, the industry will also need graphic artists. Composing graphics on a videotex system, especially the Telidon system, is easy; anyone can do it. But graphic artists bring to page design an eye educated to overall layout, color combinations, and style, which makes their skills essential in a commercial, competitive situation.

One job generated by videotex is completely new—indexing. An indexer designs the route which the user takes to find information on the system. Videotex information appears on individual pages, and the user must consult the index to find the page or sequence of pages containing the exact information sought. In a commercial system, the user pays for many pages accessed, so it is extremely important that the index be clear and well organized. If the user is confused and accesses unwanted pages, he must pay anyway. Many field trials have elicited this scenario: an apprehensive user, confused by a poorly organized index, quits the experiment, never to return. A commercial operation cannot afford to make such mistakes.

A very simple index might look like this:

```
LONDON THEATRE GUIDE
INDEX
WHAT'S NEW          1
WHAT'S BEST         2
VERDICTS            3
SHOWS BY TYPE
  CLASSICAL         4
  MUSICAL           5
  CABARET           6
```

Obviously, more complex information would require a more complex index. It is in organizing complex information that the skills of the indexer are of utmost importance. The indexer then must be able to

- structure information logically
- put himself in the place of the most stupid user
- research and test the methods chosen.

The art of the indexer is improving constantly as videotex becomes more widespread.

Some types of videotex information require the services of a content expert. At times this position is really more a role than a job and may well be filled on a part-time consultancy basis. Suppose a particular company or organization requests from an IP astronomical data in a specific area on a given date or a few pages of information about the location of geological formations indicating the presence of diamonds in a certain country. Such information is not needed every day. Depending upon the nature of the business, an IP will have on staff financial, travel, entertainment, or other experts, but the ordinary IP cannot be expected to have experts on every topic under the sun on his staff. Faced with an unusual request, he must farm out the job to the appropriate consultant or content expert.

The establishment of a videotex industry directly generates the jobs performed by IPs, umbrella IPs, indexers, and others. Without the services of the information providers, the industry would have no reason to exist. Of course, many spin-off jobs in other areas will also appear. Initially, attracting subscribers requires publicity and advertising in other media. In 1982, for example, British Telecom spent over two million dollars doing just that. Manufacturers of the videotex equipment also spent large amounts of money for the same purpose. The experience in the U.K. shows us that once videotex is in operation, the print media act as a support system. There a videotex press has emerged: four major magazines, the *Prestel Business Directory*, *Viewdata*, *TV User Guide* (a quarterly), and the *Prestel User* (something like *TV Guide*). Once again, the various media do not necessarily compete against one another. In this instance, the introduction of the new medium actually assists both the television and print media. In addition, many printed texts can be ordered by the videotex subscriber through his terminal. In fact, any printed material a company or organization cares to offer advertising its wares will receive a wider circulation if it is obtainable through videotex. Naturally all this print activity also creates jobs.

The fear that a new technology will destroy jobs is often voiced. The reality is far different. Where it is already operative in the U.K., the videotex industry is labor-intensive. For every one job destroyed, on the average three new jobs have sprung up to take its place. British Telecom, the carrier of the system, employs from 800 to 1,000 additional people as a result of its entrance into videotex technology. Approximately 4,000 people work in the service sector of the industry. That's a ratio of four service jobs created to every one technical job. And this sector will grow as companies expand their now limited services to meet the needs of new subscribers. For job seekers, the technical end is limited; the service sector is not.

SUPPORT SECTOR

Sometimes the essential role of support personnel is forgotten in discussions about jobs in a new technology. No technology, new or old, could survive for long without efficient organization. The people of the support sector are those who provide the administrative and managerial skills so necessary in building and maintaining a new industry. They are the salesmen and the technical, financial, and legal experts. Support personnel working in the videotex field fall into two main categories: those working directly in the industry and those working in peripheral areas.

If classified ads for positions in the videotex industry were run in this country as they are in Britain, we might see something like this on the screen:

ACCOUNT MANAGERS: YOUNG, DYNAMIC INDIVIDUALS REQUIRED TO ORIGINATE MAGAZINE MATERIAL AND LIAISE WITH CLIENTS. LITERACY A MUST. NO PREVIOUS INDUSTRY EXPERIENCE NECESSARY.

SALES AND MARKETING EXECUTIVE: AGGRESSIVE INDIVIDUAL WITH GOOD COMMUNICATION SKILLS, FAMILIAR WITH MARKET RESEARCH ROUTINES, FLEXIBLE--PROVEN TRACK RECORD IN MEDIA-RELATED INDUSTRY A DEFINITE ASSET.

SALES EXECUTIVE: TO HEAD ENTHUSIASTIC TEAM IN LOCAL AREA. OBJECTIVE: WIDENING MARKET FOR FINANCIAL IP.

```
VIDEOTEX PRODUCT MANAGER: TO OVERSEE PREPACKAGED
  TOURS AND FLIGHT SERVICES INFORMATION FOR TRAVEL
  IP.
```

Obviously the applicant who is conversant with the technology is at an advantage. But since the technology is new, even this requirement is frequently not met. Often the successful applicant has had managerial or sales experience in a related field—or again, no experience whatsoever. The candidate who displays great enthusiasm and radiates confidence often gets the job. New jobholders on the managerial and administrative level can learn all they need to know about the technology from on-the-job briefings. This is the great advantage of working in a new technology: competent, intelligent, creative individuals without an MBA or a commerce degree can and do get in on the ground floor. Within a short time, they become the "old hands" and rise with the industry. The most important thing is to get there first.

Where are the videotex managerial and sales positions? They are everywhere—in umbrella IP houses, hardware manufacturing firms, software firms, carrier companies, cable companies.

It is not generally realized how difficult it is to initiate a new technology in many countries at the same time. The maze of governmental regulations, technical standards, licensing requirements, and contract and pricing agreements all demand hordes of specialists with diverse skills. This is an area where one can work full or part time. A contract lawyer, for example, might decide to specialize exclusively in videotex regulations and become an acknowledged expert almost overnight. Or the same lawyer might decide to work in this field only occasionally. In addition, implementing a new technology requires careful promotion and marketing. The suppliers must be able to meet the needs of the marketplace, or the whole industry is in danger of collapse in the initial stages.

There are also jobs in international development. To understand how a new technology is set up on the international level, as it must be in today's shrinking world, we can look at the way governments are involved in promoting videotex technology. The French government, for example, has committed huge sums to developing its entire telematics industry. With a proposed twenty-five-billion-dollar investment at stake, this industry must develop export markets or it will never recoup the initial outlay. So the French government has created special branches to promote French products and services both at home and abroad.

It is worth examining these departments in detail since the services they provide and the jobs they generate are necessary in any country with a videotex industry. Intelmatique, the official marketing arm for the international promotion of French telematics products, is a multipurpose organization. It informs foreign markets about the achievements of the French industry. It offers an advisory service to foreign countries wishing to implant the technology. It interprets the needs of its customers abroad to French industry. It also aids in implementing the technology by setting up field trials or whatever else is necessary in the prospective customer's country.

Sofratev, a subsidiary of the French radio and television authorities, provides engineering services to the audio-visual marketplace outside France. It promotes and implements Antiope systems for broadcast and television operators throughout the world. Sofratev also markets the Antiope teletext system internationally. In addition, it promotes and supports the Antiope teletext standard with international technical standards groups. Issues debated by such groups include accepting or even adopting technical protocols, physical interfaces, and network transmission characteristics and parameters. Obviously, any videotex or teletext product using technical standards that are not acceptable internationally would be at a competitive disadvantage.

Antiope Videotex Systems (AVS) is the American subsidiary of Sofratev. This company is working toward a comprehensive teletext standard in the United States. It also offers a wide range of teletext services to the general public and to specific user groups. It tests the technical feasibility of the services and examines the economic parameters of commercial teletext services. AVS is also responsible for the Antiope patent in the United States and works on licensing agreements for the manufacture of the system.

Not every country which is selling or using videotex technology has organizations like Intelmatique and its subsidiaries, but the goals of these companies must be met by every nation involved in exporting or implementing the new technology. Naturally, accomplishing these goals requires the diverse skills of many people.

Telecommunications engineers handle the technical aspects. In technical matters they act as a liaison between their own staff and customers. Legal and paralegal people advise all concerned on regulations. They also negotiate and draft contracts. Joint venture licensing, which is necessary in North America, also demands the services of legal experts. Knowledge of languages and excellent communication skills are of utmost importance in all these positions.

International marketing is another job for those involved in videotex. Georges Brotman, international marketing consultant for Intelmatique, explains that marketing is first and foremost selling. To do his job, he needs the skills of a salesman. He must also have a nodding acquaintance with international law relating to the telecommunications industry. In his particular job he must be fluent in both French and English. After a sale is made, he and others also provide marketing support. They participate in conferences and continue to provide technical updating to the customer. The technical information is gleaned from in-house briefings. International marketing consultants should be conversant with the technology, but they need not necessarily be experts. It is obvious that communication skills are the foundation of this type of job as well.

These kinds of jobs in administration, marketing, and publicity will be generated by the thousands in the aftermath of the introduction of videotex to the United States.

CHAPTER SEVEN

THE OFFICE OF THE FUTURE

A technology comes into existence for many different reasons. Sometimes a device appears which has many users if its potential consumers can be convinced that they need it. This is the case of a technology like videotex. Its possible applications are enormous and really only limited by the imagination of its users. This type of technological development is known as a technology-push change, meaning that the existence of the technology pushes us to find a use for it. The new technology is adopted because it allows a whole new class of activities to be undertaken. Such technological developments have a long and honorable history and have produced such things as the telephone, the camera, and more recently, the microchip.

More common, however, is the technology-pull type of innovation: It answers a real need in the marketplace by offering a way to perform current activities more efficiently or more cheaply. Technology-pull innovations include most of the inventions that were instrumental in the rapid development of the Industrial Revolution: the cotton gin, the steam engine, and the mechanized loom. The office of the future, variously known as the automated or electronically integrated office, falls into the technology-pull category. The office of the future is technology's solution to the contemporary office problems of rising costs and declining productivity.

THE AUTOMATED OFFICE

Although introducing computers and computer-related technologies into the home is indeed a great innovation, the computer has long since become a standard feature in all large organizations. In fact, in the world of business, the computer, which only a few years ago was the apex of electronic sophistication and a complete mystery to the average person, has already been superseded by the advanced communication technologies of the office of the future. Automated office technology provides the means to connect all business machines, telephones, computers, typewriters, photocopiers, and the like with each other and with the outside world. It provides the basis for an expanded electronic information network whose complex grid will eventually link the entire world. It is not just one technology, but, rather, the convergence of many, including mainframe, mini- and microcomputers, telephones, satellites, lasers, and fiber optics.

Although still in its earliest stages of implementation, office automation is occurring at breakneck speed. It has already created a multibillion-dollar world market for its products and, incidentally, enormous opportunities for acute individuals. In a special report analyzing world market trends, researchers for International Data Corporation predicted a 33 percent compounded growth rate for the 1979–85 period. Sometime in this decade, the annual market for office automation products in North America will reach an astounding $300 billion. By 1990, the United States market for work stations, groups of electronic office machines, alone will absorb twenty-eight million units. These astronomical figures represent only sales of hardware, the actual machinery of the integrated office. As our previous investigations have shown, hardware manufacture represents only a small portion of the actual total market in any high-tech communication industry. A look at subsidiary or spin-off industry, along with support and entrepreneurial sectors, shows not only real dollar amounts which are considerably higher but also a phalanx of hidden employment opportunities.

The office of the future, though, is different from the other industries we have seen for a number of reasons. For one, many technologically viable forms of the automated office now exist; no one knows which alternative will become the standard or, indeed, if there will be any standard. To further complicate the issue, most of the potentially standard technologies are already gigantic industries. The scope of the office automation industry is also unbelievably vast. During the past several decades, the administration of every sector in the economy has expanded to the point where more than one half the entire labor force now finds employment in some type of office. The coming of the office of the future will transform the working lives of millions of people. In addition, the widespread implementation of the electronically integrated office throughout this decade and well into the 1990's will firmly establish as reality the much-heralded Information Revolution, which, until now, has been the stuff of science fiction. This chapter will describe the new technology, its applications, and alternate technical forms, as well as the numerous job openings soon to be available in the hardware industries. The following chapter is a more speculative discussion of the potential effects of the Information Revolution, which no one can predict with exactitude, on society in general and employment in particular.

The exact form the office of the future will take is unknown. What is known, though, is that some form of the automated office will become a reality in the 1990's. The need for the automated office, although most of us are unaware of it, is very great. The automated office provides a solution to the twin problems of runaway costs and declining productivity, which bedevil modern administrative organization. The modern office, for all its shiny glass and chrome veneer, is, in fact, antiquated.

CRISIS IN THE OFFICE: THE PROBLEM

Harried secretaries type furiously while answering ever-ringing telephones. With a deafening clackety-clack gigantic typing pools churn out reams of paper. Busy cigar-chomping executives talk endlessly on several telephones while dictating letters to blonde bombshells. This Marx Brothers version of office activities, though stereotyped and dated, is fundamentally accurate because it emphasizes the office's basic function: Communication is the name of the game. More and more, however, the modern office is not fulfilling its most essential role.

The modern business office has only one product: information. It performs only five basic tasks: producing, communicating, collecting, processing, and storing information. When we say that office productivity is declining, we mean that one or more and probably all of these information operations are in jeopardy. Each task is, roughly speaking, facilitated by a corresponding machine: the typewriter for producing information, the telephone for communicating it, the computer for collecting and processing it, and file systems for storing it. But although desks may be jammed with sophisticated equipment, employees find it hard to locate and communicate essential data. Executives complain that while paper flows in rivers down their company's corridors, they still lack fundamental information or find it too late. Important decisions are delayed or never made as a consequence. Worse yet, the wrong decisions are made on the basis of inadequate, out-of-date, or incorrect information. Perhaps in the past all this was tolerable, but in today's global business world, the rapid location and communication of information is an absolute necessity. Threatened by the specter of fierce competition from foreign shores, American corporations and institutions must cut costs and boost productivity or face extinction in the marketplace.

The familiar business office is in full crisis. Creaking along like a stone wheel in an age of space flights, it needs help. Efficiency has been declining for decades; productivity is falling. As a result, costs have skyrocketed as business has tried to increase output by hiring even more workers. Since the 1950's, the percentage of employees engaged in some type of office work has been steadily mounting; today, their salaries represent some 92 percent of all direct office costs. In the past ten years, management has invested an average of $24,000 per factory or production worker, increasing productivity by about 84 percent. During the same ten-year period, business invested about $3,000 per office worker, resulting in a dismal 4 percent increase in productivity. In fact, according to New York management consultants Booz, Allen & Hamilton, Inc., since a high in 1978, productivity in the office has even fallen off a few percentage points.

Clearly the situation is becoming desperate, for within the next few years office costs will double yet again. The patchwork solution of the past is no longer possible. In today's economy, the cost of adding more workers is simply prohibitive. Indeed, it is obvious to all that an increase in productivity is mandatory. Information must be available, accurate, and current. Communication must be easy and rapid. Fortunately electronic technology is making the gargantuan feat of improving office productivity a reality. And this technology promises even greater feats in the near future. For those who are prepared in advance, the new technology will open new career paths and great opportunities.

THE DECEPTIVE SOLUTION: WORD PROCESSING

During the 1970's, as the economy went into a decline, poor office productivity, which had been tolerated and even encouraged in times of prosperity, suddenly became The Problem in the race to cut costs and bolster sagging profits. As luck would have it, the appearance, at about the same time, of the mighty microchip made possible a marvelous machine which at first seemed the answer to every prayer and the final solution to the productivity issue. The word processor, a cross between a typewriter and a computer, can, after all, triple the output of a single typist.

A word processor is a computer programmed to perform multiple operations on textual materials. In its most basic format a word processor (WP) consists of a video display screen, a keyboard, and a computer. An operator can write, revise, and edit a text by displaying and manipulating it on the screen through commands given at the keyboard. Since the text appears on the video screen, there is no paper at all involved in this process. By pushing a few buttons, the operator can perform all the operations of a sophisticated typewriter: justifying margins, centering headings, and changing type styles. The WP is also a true innovation, however, for an operator can command the machine to skip lines; insert, delete, or store blocks of text in its memory; transpose parts of the text; insert one document into another; locate a word; jump to a new page or phrase; repaginate—in short, perform editorial functions automatically. The final text can be stored in the computer on mag media—mag(netic) tape or disk. Connecting the machine to a good printer results in hard copy of high quality in a fraction of the time it would take a typist to produce it.

The uses of the WP are many and still expanding, as more sophisticated machines come onto the market. Phenomenal increases in clerical productivity are now a matter of course. Consider, for example, the typical business letter. A manager dictates a letter, which is then typed and submitted for proofreading. If there is a mistake in it, or if the manager wants to add or delete some material, the letter must be retyped and reread. Eventually the letter is mailed, but it may take three days to a week to even get out of the building. This tedious process occurs hundreds of thousands of times a day across the continent, wasting valuable work time. Word processing eliminates the whole operation.

The WP is a heaven-sent gift in the production of long documents requiring a good deal of revision and editing. Because WPs can store text in memory, law offices no longer need to retype standard legal documents over and over: the client is served in minutes instead of days. A name, place, or date incorrectly cited in a two-hundred-page document used to involve manually searching through the documents, correcting the error, and sometimes cutting and stapling the whole document, but not with a WP. WPs can produce personalized mass mailings, keep addresses and phone numbers up-to-date, do payrolls, record appointments and meetings on an electronic calendar, and generally take the tedium out of routine office procedures.

Second-generation communicating word processors (CWPs) can "talk" to other computers through telephone and satellite links. CWPs can call up information from remote data bases, distribute documents and messages to multiple users, act as a file for personal schedules, and much more. In addition, any microcomputer can easily access material held by a CWP.

Increasing typing productivity costs money, of course. The average WP costs from $7,000 to $15,000. The price of the machine naturally depends upon its quality and functions. IBM's popular Displaywriter, for example, selling at about $15,000, boasts a "dictionary" which can check the spelling of over 55,000 English words. Add to that the cost of peripherals, such as a printer at $10,000, and one begins to see the magnitude of the burgeoning office equipment market. Hardheaded business people apply a strict cost-benefit analysis: a 10 percent increase in the productivity of a $30,000-a-year office worker justifies a $15,000 outlay for equipment with a five-year useful life. WPs pass the test every time.

It is readily understandable how these marvelous machines, able to perform miracles with written texts, could have initially appeared as the solution to poor office productivity. Typing production is easily quantifiable; documents and letters are tangible evidence that WPs produce more quickly, efficiently, and in greater volume than previous machines. That WPs are cost-effective is immediately obvious. But as the more astute business people soon came to realize, improving structured office work attacks only part of a much larger problem. Clerical support of all types accounts for about 34 percent of total office costs, secretary-typists for a mere 6 percent. Typing itself, taking up only about 20 percent of the secretary's working hours, represents a minute 1.2 percent of the global total. While it is true that this "minute" 1.2 percent costs American businesses $4.4 billion a year, clearly, automating the office has to go far beyond typing to have any real effect on productivity.

Automation has already taken place throughout the rest of the clerical sector, as inventory, accounting, and the like are generally computer-based. So improving unstructured office work, representing a whopping 66 percent of all office costs, must be the real aim of any serious effort to improve overall productivity. Managers and professionals, the two groups performing most unstructured work, spend up to 95 percent of their working days engaged in some kind of communication, oral or written. In fact, white-collar workers from all sectors spend about 40 percent of their time in oral communication. Automating the office means improving communication.

THE ELECTRONICALLY INTEGRATED OFFICE: THE SOLUTION

Today many offices have what amounts to a hodgepodge of specialized equipment. Almost every office, large or small, has filing cabinets, telephones, typewriters, and calculating machines to perform the basic information operations of the office. These electromechanical gadgets may look streamlined and modern, but each and every one was already a fixture in the office of fifty or even a hundred years ago. Through advances in electronics a wide range of new labor-saving devices has appeared in the office. However, this new electronic capability has arrived in bits and pieces of specialized equipment, creating in most offices what can be termed the electronic hodgepodge: typewriters, word processors, computer terminals, printers, copiers, facsimile devices, and more. Although designed to save time, these machines often cause confusion, unnecessary duplication of effort, and ironically, an actual decline in productivity.

The office of the future aims to reduce wasted effort by enhancing communication, the most essential function of the office. In the automated office all business machines will be electronically connected with one another and with the outside world. Potential savings in time and labor costs will be enormous, as just a few examples make clear.

In the contemporary hodgepodge office, a typist produces a document, walks to the copier, waits for the duplicates, and returns to file them. If the WP, printer, and filing system were interfaced, or able to communicate electronically, the typist could produce the documents more quickly and the rest of the operation could take place instantaneously as a single action. The typist would be free for other tasks. Or a manager calls the vice-president of her division and, after four or five tries, gets through. Then she writes a memo confirming and recording the conversation. If the telephone, the WP, and the filing system were interfaced, the call, message notation, and its storage could take place simultaneously. Today's sophisticated telephones can even eliminate the ever popular game of "telephone tag," which now has office workers leaving messages and calling each other back, always unsuccessfully, sometimes for days at a time. Both the manager and her secretary would save valuable time.

Exactly how much time could be saved is difficult to determine, since productivity in communication is hard to quantify. Nonclericals may be so highly paid, however, that even small time savings can mean great returns on investments in costly equipment. Some estimates put time lost to shadow functions, useless and unforeseen activities associated with accomplishing any task, as high as 8 to 10 percent of every working day. Only 28 percent of business calls reach the intended recipient, making attempts at phone communication a major shadow function, though it is a vital activity for nonclerical personnel. Receiving calls wastes time, too, for the recipient must put aside current work and attend to the caller. Such interruptions cause wait and recycle periods before the recipient can resume previous work; they are so common in unstructured office work that it has been described as "interrupt-driven." At least sixty minutes a day is lost to interruptions; the quality of work no doubt suffers as well. Diminishing time loss among nonclericals means big savings: even a gain of two hours a day could result in $62.5 billion a year, or $15,000 a year per employee.

METHODS OF OFFICE INTEGRATION

Automating the office can almost assuredly effect dramatic jumps in productivity, especially among highly paid nonclericals. Business is eager to automate, but confused as to how to go about it. The established hardware giants, anxious to be a part of the new industry, have proposed a bewildering variety of different technical strategies; all are designed to increase productivity by interconnecting office machines and outside telecommunications networks. Essentially the chaos resolves itself into two fundamentally different methods, both extensions of methods used in data processing, namely the distributed versus the centralized approach. The centralized approach is to boost the electronic intelligence of a basic office machine, either the computer or the telephone, and interface all the other machines with it as "dumb" terminals, which are unintelligent and incapable of performing any task alone. The distributed, or decentralized, approach is to boost the electronic intelligence of many machines and to interface them through some kind of internal communication network, either a passive cable with no intelligence of its own or an enhanced telephone line. The following diagram presents the methods, the means, and some sample proponents of each schematically.

Centralized Distribution	
means	*sample suppliers*
mainframe computers	IBM
telephone (super PBX)	Northern Telecom, Rolm

Decentralized Distribution	
means	*sample suppliers*
cable or data bus	Wang, Datapoint, Xerox
telephone line	Bell System (AT&T)

FIGURE 7.1

A more detailed discussion of each of these approaches follows in sections evaluating the major office machines. As the leap from plain old typing to word processing shows, electronics has radically altered the capabilities of one primary office machine, the typewriter; the computer, the telephone, and filing systems have undergone changes as well. Each will have an expanded role in the office of the future.

The situation in the nascent automated office industry is so chaotic that many experts compare it to the early days of the automobile industry. Back then, competing firms made cars powered by steam, electricity, and the internal combustion engine. Of course, internal combustion won out, and the others became quaint relics of a bygone era. This may well be the case in office integration. On the other hand, the various methods of office integration could continue to coexist, each attracting its share of adherents. For some time now, futurologists have predicted that the age of homogeneous America is over and diversity is the new password. All automobiles did not have to use the internal combustion engine. Perhaps something in the spirit of the age encouraged a standard product. Perhaps just as powerful a spirit will discourage standardization in our own age. Technically there is no overwhelming reason why there should be a standard.

THE COMPUTER. The computer or data processor performs two of the five basic tasks of the office, collecting and processing information. Throughout its short history, the computer has had undoubted success in automating the significant portion of clerical work generated by these two tasks. Mainframe manufacturers propose to extend this success by making the computer the central hub of a fully automated office. The computer's history, though, reveals developmental difficulties leading indirectly to an alternate automation strategy, the decentralized data bus system, sometimes called the local area network (LAN). A brief look at the computer's past shows how this happened.

In the early days, only large corporations could afford to buy or lease the huge mainframes. Typically the corporation established a separate data-processing division to generate programs for specific jobs and oversee the whole operation. In effect, the computer, supported by data-processing personnel, who alone had the arcane technical knowledge necessary to understand and control the system, constituted a kind of centralized office.

Computers speed things up considerably, but as we all know by now, they can also cause problems. In the not-so-distant past, using a computer meant using batch processing. After laboriously punching an IBM card for every single line entry, the operator had to wait for time on the computer before putting the job on. Batch processing was so slow that another method emerged, multiprogramming, which allowed several jobs to run at the same time, but which also created another set of equally irritating problems.

Eventually computer manufacturers developed interactive processing. Eliminating cards, this method allows the operator to enter information directly through a keyboard. Hundreds of terminals connected to the mainframe can communicate at the same time and receive information immediately; thus it is also called distributed data processing. Mainframes designed for batch processing, such as the famous IBM 360, do not tolerate the new method well. Each new terminal added slows down the computer until, suddenly and without warning, it "goes down." When a computer "crashes," all work is taken off, causing frustrating delays until the machine is back in action.

The demand for interactive processing grew until finally new machines were built specifically for the new distributed data processing. These machines still crash, but not often. While mainframe manufacturers were wrestling with new designs, microcomputers burst onto the market and provided a completely different solution. Small, inexpensive, self-contained microcomputers are programmed for specific tasks, then positioned around the office in strategic locations. A microcomputer might be programmed as a word processor or to handle accounts receivable and payable, payrolls, personnel records, or a myriad of other things. The hitch is, though, that self-contained microcomputers, sometimes called standalones, cannot communicate with each other except by the use of costly techniques, and just like the mainframes when overburdened, they can also go down.

One way or another, though, distributed data processing is here to stay. For office automation the next step in the sequence is to permit all office machines, many performing text functions, to intercommunicate. As we've already seen, this can be done through the local area network (LAN) technique or through the mainframe.

Both systems do the same thing by different methods. The LAN integrates all office machines by means of a special cable, usually coaxial, which functions as a kind of data vehicle, or data bus. Each office machine is given a special interface which allows it to connect with the cable and communicate with all the other machines on the network. There is no central controller, and the data bus itself is really only a passive carrier. The information-handling power lies in each individual machine on the network. Through connections called gateways, the whole system communicates with the outside world via ordinary telephone lines, satellites, or other common carriers. One such system, the Xerox Ethernet System, looks like the illustration on page 164.

Routing the whole system through a mainframe computer accomplishes roughly the same thing, except individual machines in the system need not be intelligent. Depending upon the power of the mainframe, an almost limitless number of terminals can join the system; established data processing departments keep the whole network running smoothly. Computers, too, can communicate with the outside world through gateways.

Each system has its advantages and disadvantages. Proponents of the data bus approach say it combines the advantages of mainframes and micros, yet since the bus directs the flow of communication, no computer power is wasted directing traffic. The bus carries information almost instantaneously in both directions, so the system has the speed and efficiency of a mainframe even though its elements are physically dispersed like micros. As the bus reaches its load capacity, another bus can be added and another and another.

True, say mainframe manufacturers, but we offer the same. Mainframes tolerate hundreds of terminals; if even that is not enough, upgrade the computer. To help limit the number of terminals, we offer subsystems within the office which depend on the mainframe for central storage and extended processing capability. In addition, say mainframe manufacturers and data processing professionals, we see some fundamental problems in the data bus setup. Both micros and minis tolerate only a small number of auxiliary terminals. If you want data to be shared by many

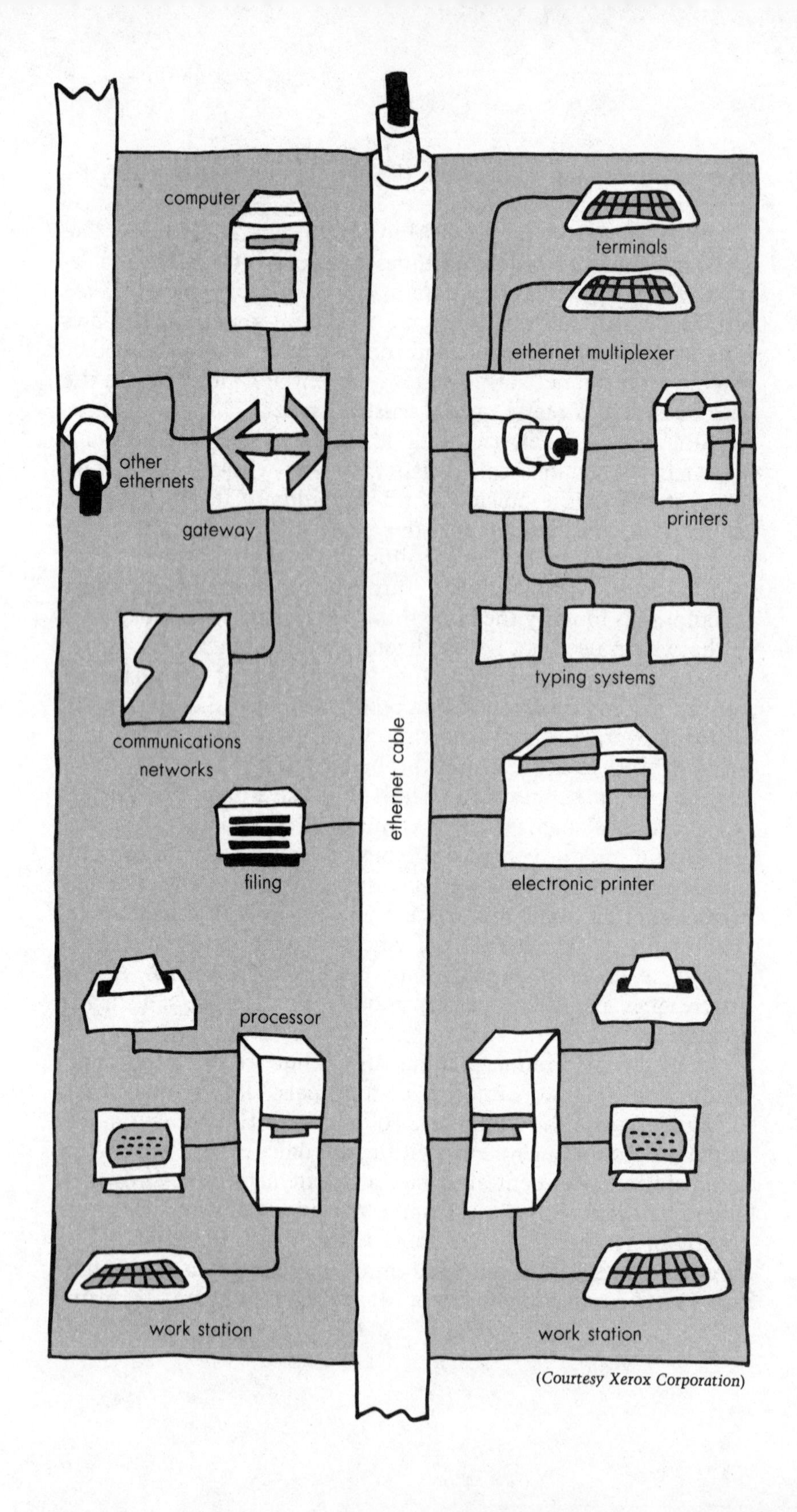

(Courtesy Xerox Corporation)

machines, that's inconvenient and expensive because you'll have to keep buying more minis. In the bus model minis and micros can be distributed in remote sites throughout the company or even throughout the country. But what about security? What information is where?. Who has access to this data? Who's controlling the flow? Who even knows where it is? What about data integrity? If you distribute data all over the place, how do you keep it all current? It's likely that, thinking the information in the micro is up to date, someone is going to make a costly mistake.

Over the short term, mainframe manufacturers will most probably capture a significant share of the office automation market. Large multinational corporations, most of whom already have mainframes and supporting data processing divisions with extensive computer experience, will be the first to implement the office of the future. Multinationals present a most lucrative market.

In the longer term, as the office of the future comes to organizations of all sizes, it is likely that the data bus approach will win the lion's share of the market. Capital costs associated with the mainframe route are astronomical and out of the reach of smaller companies. The bus system is relatively inexpensive; minis and micros are far cheaper than mainframes. Buying a mainframe means total commitment to one approach; with the bus system the purchaser is still free to change in midstream. And the purchaser does not have to buy everything at once but can proceed at a convenient pace. Further, by offering interconnection with existing mainframes, bus systems manufacturers are in a position to compete directly with mainframe producers.

Whichever system prevails, it will assure one of the major advantages of the integrated office: the sharing of resources among its various departments. In the electronic hodgepodge office, a $10,000 printer in the personnel department, for example, may print out payroll checks twice a month, organize pension and social security data, prepare personnel dossiers on a routine basis, and still be idle for significant periods of time. In the integrated office the printer can be programmed to produce for other departments during its free hours. All material arrives electronically, so

FIGURE 7.2

it makes no difference where the machine is actually located. Thus time is saved and the productivity of both man and machine will rise.

THE TELEPHONE. Like indoor plumbing, electricity, and hot and cold running water, considered luxuries not so very long ago, the telephone is a modern necessity, found in 95 percent of American homes. Many of us have two or three, some in different colors to fit the decor. Running a business in North America without a telephone is unthinkable. But while the majority of subscribers take the telephone for granted, big things have been happening to it.

The telephone performs the basic office task of communicating, but today's telephone is no longer the simple voice-message carrier of yesteryear. Thanks once again to the microprocessor's marvels of miniaturization, the telephone now handles voice, data, text, and even video. The enhancement of business telephone service has resulted in an instrument which can perform a multitude of operations to save time and increase productivity.

If the president of the company is doing a slow burn because his calls have failed to get through just once too often, he can install a phone system featuring priority queuing, programmed so his calls will always get through. Some telephone systems offer smart-routing, which automatically chooses a WATS or a long-distance line, depending on which costs the least; some have hands-free service so the client never has to pick up a receiver; others, automatic call back, or camping-on, which automatically keeps ringing a number until it answers. Some systems will keep calling the last number dialed; others dial preprogrammed numbers at the push of a button. Some feature conference calling or paging capability. With automatic call-forwarding, a call will automatically be forwarded to another number. Station message detail recording (SMDR) is a feature popular with management. A traffic analysis tape keeps track of who called, who received the call, the length of the call, the time, the kind of line used, and the cost. Cost is a constant headache. For many large organizations, telephone costs come third in the overall budget after salaries and rent; for others, such as stockbrokers, telephone costs may even be second. SMDR can be an important option.

Northern Telecom's Displayphone is today the most advanced

on the market. It is really an advanced microcomputer terminal which integrates the telephone and the computer into one desk-top unit. This "telephone" can be connected to almost any line—computer, telephone, or other; it has a video screen, a touch-sensitive key pad, and also a handset (a receiver to the rest of us). In addition to the capabilities described above, this machine can keep your personal schedule in order and even call up the schedules of other people with Displayphones and slot in meetings when all are free. When you go back to your office after lunch, presto! another meeting has appeared on tomorrow's agenda. The phone can also store messages and handle electronic mail.

Because of these new capabilities, the business telephone may well play the role of supercontroller in the office of the future. The lowly telephone may just become the central hub of the integrated office. The modern story of the telephone really begins with the 1968 Carterfone decision of the United States Supreme Court, which deregulated part of the telephone company's (telco) business by allowing subscribers to install their own equipment on telco lines. Up until then, equipment supply had also been a telco monopoly. The 1980 Canadian Radio-Television and Telecommunications Commission (CRTC) decision did essentially the same thing for parts of Canada's industry. The Supreme Court decision alone would have aided in the creation of a tremendous new private equipment industry, called interconnect, but, by chance, that decision came at just about the same time as the invention of the microchip. These two almost simultaneous events made the common telephone into an even more extraordinarily powerful tool of communication. The new private equipment companies had to offer very special products in order to lure customers away from Bell; the microchip gave them the chance to design and develop the telephone in ways they could never have imagined in those beginning days.

Sometimes we are slow to understand when a great opportunity presents itself. Just as the microcomputer was originally targeted as a gadget for the home hobbyist before people began to appreciate its applications in the far more lucrative business market, so, too, did interconnect stumble before it walked. Although interconnect got its start selling fancy telephones to the home market, soon interconnect companies realized the massive potential of this first chink in the monopoly armor of Ma Bell. They began selling key telephones, the sets with push buttons used in business to handle several lines. Then interconnect offered private branch exchange (PBX) systems, entire internal phone networks. The PBX operates by means of a switching device which connects the private network with standard telephone lines. Companies that buy a PBX system are said to have "bought a switch."

In the early days of telephone science, individual phones connected directly with the telephone company's central exchange. As the number of subscribers grew, so did the number of wires; old photographs show city streets looking like giant arcades, their roofs blackened by wire. The unsightliness eventually forced another solution: a PBX, or switchboard, transferred calls within a building, while outgoing calls reached the telco's exchange via trunk lines. A modern PBX performs essentially the same function it always has, but the Carterfone decision opened the doors to competition and a barrage of innovations. PBXs are now "enhanced."

Microchip technology complicated and at the same time extended the capabilities of these switches and thus the telephone. In the past, data and voice communications were two separate and distinct operations. Now all that has changed as the boundaries separating the two have irretrievably blurred. Computers can "talk" to other computers and can do so over telephone lines; the human voice can be translated into the digital language of computers. Both modes can use the same line at the same time. In 1975, the first computer-based software-programmable PBXs appeared, designed for the voice, that is the telephone, market, to aid in processing and tracking telephone calls; these machines were for analog, or voice, switching. Most voice transmissions are in the analog mode, while most data transmissions are in the digital mode; they are mutually unintelligible. Then Northern Telecom put the first digital PBX to integrate voice and data switching onto the market at the beginning of the 1980's. At one end, such a machine translates signals sent in the analog mode into digital, so that both can occupy the same line; then it translates the signal originally sent by voice back into analog at the receiving end. With all data and voice transmissions now in the digital mode, all office machines are poised to communicate with each other. This versatile switch is really a communication infrastructure: the PBX can become the supercontroller of the integrated office.

Companies, including Bell, Rolm, and Northern Telecom, already offer super PBXs as central controllers in their versions of the office of the future. It is also worth remarking that Datapoint's LAN includes a sophisticated third-generation PBX, making it the only fully integrated office system on the market today. A most advanced concept of the super PBX was announced by Exxon Corporation's InteCom subsidiary and called the IBX. This system has all the usual PBX features coupled with an ability to convert protocols and formats so that dissimilar office machines can communicate without modification. AT&T's favored entry in the office automation sweepstakes, System 85, is at the moment Bell's best bet. This powerful PBX, supporting several minicomputers, can also fully integrate the office; it even monitors building security, right down to checking passes, and programs air conditioning and heating. Trade publications also report serious research and development by IBM (Project Enterprise), Datapoint (Project Evergreen), and AT&T (Project Antelope) on an even more superior super PBX.

The super PBX as central controller of the office of the future is just one telephone-based automation strategy. AT&T, the biggest company in the world and prime mover in telephony, cannot be expected to let the immensely lucrative office automation market slip through its fingers. Bell did, in fact, propose a distributed approach to office automation, the Advanced Communication System (ACS). ACS was to enhance the already considerable intelligence present in the public telephone network to allow dissimilar machines to communicate with each other. Though at one time widely acclaimed by Bell's veritable army of PR personnel, lately ACS seems to have died a quiet death. Since it does not exist in reality and may never appear at all, assessing its impact is impossible. The almost incomprehensible might of AT&T alone, however, makes that company a most significant player in the office automation arena, whatever its current difficulties there.

Although a recent Justice Department settlement divested AT&T of its local telcos, this will allow the company to devote its enormous resources more single-mindedly toward developing the most advanced technology. Left intact by the settlement were its mind-boggling infrastructure, its Long Lines long-distance network, its gigantic Western Electric subsidiary, and last but not least Bell Labs, originator of some of the most provocative new technology of the century. AT&T could easily become the most formidable competitor in the office integration arena.

FILING SYSTEMS. The office filing system is used for storing and retrieving information. Increasingly files fill wall after wall and drawer after drawer in many offices. There is a most pressing need in today's office for a system which is organized, easily accessible to many people at the same time, and secure, and which takes up little expensive office space. An effective filing and retrieval system could generate savings far in excess of those offered by WPs. Unfortunately such a wonderful device awaits invention. Even the automated office of the future has not solved this problem.

At the moment, storage and retrieval are accomplished through three basic media: hard copy, microform, and magnetic. Of these, hard copy is by far the most prevalent. Vertical and lateral files are a feature of almost every office and will continue to be so for quite some time to come. Microform comes in two common varieties: microfilm and microfiche. Microfilm reduces the need for space while increasing security, to be sure, but it needs processing and a special viewer. Indexing and retrieving documents present difficulties, and updating a photograph is impossible. Nevertheless, this industry experienced a 20 percent growth rate over the past five years. Microfiche cards are particularly useful for unit records, such as medical, legal, and personnel files. The cards can be easily distributed, and the readers are inexpensive. Updating records is frequently necessary, so microfilm jackets or covers with updated information are used. All microfilm is relatively expensive; typically a business must accumulate at least seven years' worth of hard copy to justify its cost.

Now the industry seems poised to overcome these disadvantages with the latest twist: the merger of microcomputer technology with existing micrographic techniques to create computer-assisted retrieval (CAR) systems. CAR offers faster, easier retrieval. The intelligence of the microcomputer allows documents to be filmed and indexed at random. The computer keeps track of the location of the films and directs the microfilm readers to the right frame. Most machines have optional printers. CAR systems currently available include an integrated system which acts as a peripheral to the user's mainframe, such as Kodak's IMT-150, and standalones which operate independently, like those from 3M and Bell & Howell. Although the use of CAR is growing, this type of system is not all-purpose. It is best suited to transaction documents, such as accounts payable and receivable; for general use, it must function with the other storage media mentioned already.

The third type of filing system is magnetic media, used in conjunction with a computer and serving as a combination external computer memory and programming device. Like the CAR systems, they are extremely compact, and they are even more accessible to the computer. The major types of magnetic media offered by integrated office equipment suppliers include the floppy disk and diskette. First developed about ten years ago, floppy disks are 8-inch-in-diameter mylar plastic sheets on which information is recorded in circular tracks. Floppy diskettes, a 1978 innovation, are 5¼ inches in diameter. Both resemble phonograph records encased in cardboard holders. Capacity has stabilized at around 500,000 characters for a disk and 250,000 for a diskette. However, new advances in recording head materials, drive speeds, and packaging promise storage of two to three million characters for floppies. Disks and the drives into which they are inserted are really computer input devices. A program for a computer is loaded onto a disk, which is then inserted into a disk-operating system. Floppies themselves sell for only a few dollars, but loaded with intricate computer programs, their cost jumps into the hundreds. The disks are used both to store data and to program the computer. Floppies offer enormous space savings, indexing is no problem, and they can even be tossed into the mail. Any type of computer memory, however, has limits. Floppies are best used for extremely interactive files which must be updated frequently. Using magnetic media for archival material is a kind of technological overkill, a misuse of a valuable tool.

The ideal filing system has not yet been invented. It should be equally useful for interactive and archival data; it should be cheap, easily accessible, and able to produce hard copy on demand. In the immediate future, business will continue to use a combination of the three types of filing systems.

THE OFFICE AUTOMATION INDUSTRY: JOBS

The office of the future is no longer a dream. The computer, communications, and office equipment industries have converged into one gigantic industry. Today the office systems industry is on the verge of widespread implementation. That implementation will take place throughout the entire spectrum of North American administrative organization; it will affect every single United States business, from the largest to the smallest, as well as all levels of government and every kind of institution, educational, medical, ecclesiastic, and otherwise. Implementing the office of the future on so vast a scale will take the rest of this century to complete. Analysts predict that soon the North American market for office automation products will have reached a stupendous $375 billion a year, with the relatively small Canadian market representing $15 billion of that total. The birth of a new industry of such enormous proportions signals outstanding opportunities for the qualified job seeker.

The prime market for office automation products today is the 12 percent of United States businesses employing over fifty people. Those firms currently employ nearly 75 percent of the entire American labor force. The federal government, the nation's largest single employer, is another major market, along with the bigger state and municipal administrative units. Initially, most analysts agree, electronic office integration will come to Fortune 500 companies or their governmental equivalents, with the very top fifty firms in the vanguard.

Already well advanced into automation, only these largest organizations have the technical expertise and experience to make a smooth and rapid transition to the new technology. Further, these firms and institutions have the most immediate need for electronic networking. In businesses where the main activity is service, such as banking and insurance, office automation is already a fact of life. In these institutions office automation is like factory automation: it speeds up production of the product and lowers its cost. For some industries, especially those facing the double whammy of a mounting crisis in productivity coupled with expanding foreign competition, notably from Japan, in markets hitherto considered "safe," office automation may be the only solution left.

After the very largest organizations, possibly the smallest businesses, at the other end of the economic scale, will be the next to embrace electronic office systems. Many professionals, accountants, lawyers, dentists, doctors, and the like are now streamlining their business operations with microcomputer technology. Positive experience with technical innovation may well encourage them into full-fledged office integration. The greater number of middle-sized companies will most probably be the last to join the electronic fraternity. It is worth noting as well that according to a recent study by a consortium of German universities, a full 20 to 40 percent of all existing communication modes could be replaced by implementing the automated office.

Since implementing the office of the future will continue well into the twenty-first century, activity in the manufacturing sector should be intense for quite some time to come. Demand for the first wave of office integration products will peak sometime in the mid-1990's as the largest firms and institutions satisfy their initial requirements. But demand for product should nevertheless remain strong when the major market becomes the many latecomers to office integration. By this time, as well, initial users will be clamoring for increasingly sophisticated products, since for some enterprises technical advances will be imperative if they are to maintain the competitive edge. The outlook for the manufacturing sector could not be better.

Because office automation may take a variety of technical forms, its manufacturing sector is far more complex than most. For, as should be clear by now, office automation is really an umbrella industry comprising many presently established individual industries. The manufacturing sector will generate thousands of jobs. For the sake of simplicity, the discussion of the manufacturing sector is divided into direct equipment manufacture and subsidiary or spin-off manufacture. Spread out among a score of large and growing industries, direct equipment manufacture is already a gigantic sector, while subsidiary manufacture represents a diverse and expanding field of almost unlimited entrepreneurial potential.

DIRECT EQUIPMENT MANUFACTURE

A list of direct equipment manufacturers in the automated office industry would read like the corporate *Who's Who* of America. The lure of huge profits in a stupendous new industry pits corporate titans like IBM and AT&T against each other on an immense field of battle. The size of the battlefield is almost beyond comprehension. It is today dominated by a handful of Fortune 500 companies whose combined revenues stagger the mind. Nor is combat limited to already established giants: the numerous technological alternatives invite new entrants into the fray.

In order to facilitate the discussion of job opportunities in this complex new industry, this section will be confined to three basic areas: *computers,* including mainframes, mini-, and microcomputers, as well as their respective software components; *local area networks* (LANs); and the *telephone alternatives* of the interconnect industry and the vast public system. Other related industries come under the heading of subsidiary manufacture.

All the products of these varied industries must be researched, developed, manufactured, marketed, distributed, and supported. Each of these manufacturing functions generates employment. Broadly speaking, technically trained professionals find employment in R & D and oversee manufacture, while individuals with business skills and an understanding of their particular industry work in the marketing, distributing, support, and general administrative areas. Since many of the technical and support positions are the same as those already encountered in the discussions of other industries, this section examines only employment not before mentioned. In providing a guide through the complexities of present-day direct equipment manufacture, this section presents an overview of each industry, its present scope, and its future prospects as office automation becomes reality.

COMPUTERS. Until recently the computer industry, now often known as the information processing industry, produced only one product: the big mainframes, today selling for between two and four million dollars each, found in the larger businesses and institutions. Within the last few years, however, the market has grown and fragmented with the advent of minicomputers and business microcomputers. The terms "minis" and "superminis" here refer to machines priced from $25,000 to $250,000; and "microcomputers" to machines which usually sell for less than $10,000 but whose increasing sophistication in the business environment may push the price up to $25,000.

Mention mainframes and someone is likely to say IBM—International Business Machines. IBM has dominated the computer industry for almost its entire history. In fact, the whole mainframe industry defines itself in terms of IBM's machines. Aside from IBM itself, the industry consists of the so-called plug-compatible manufacturers, producing machines able to stand in for those of IBM, and the five companies, known as the "Bunch," whose machines are not compatible with IBM's.

Presently mainframes account for about 60 percent of the United States computer market, but their share is reputedly shrinking. According to the Framingham, Massachusetts, firm International Data Corporation, the mainframe share of the computer market may shrink to less than 40 percent within the next couple of years. This is bad news for mainframe manufacturers, of course, but all is far from lost. Certainly information processing as an industry seems poised for a major shift in market patterns. Many companies, though, having seen the handwriting on the wall a few years ago, have already diversified into other product lines and are thus now in a relatively strong position, ready to move into other facets of the market.

In any case, it is hard to work up much pity for IBM, whose annual worldwide sales run into the region of a healthy $34.4 billion with profits of $4.4 billion, making it the most profitable United States industrial company. IBM commands some 40 percent of the worldwide market for computing equipment and produces about two thirds of all mainframe computers. In just about every one of the 130-odd countries where it does business, IBM is the leading computer company. While IBM may bemoan the loss of its traditional 90 percent market share of just a few years ago, the world market is constantly expanding, and more competition should spur the company to scale greater heights. IBM is not about to give up the ghost. In fact, it recently launched its most advanced mainframe design to date, the 3081-K, and made a splashy and successful entry into the microcomputer market.

Backed by its formidable reputation for excellence and its experienced sales and support force, IBM will continue to be a major presence in the office automation arena.

Plug-compatible manufacturers (PCMs), such as Amdahl, Memorex, National Storage, and National Advanced Systems, a subdivision of National Semiconductor, produce equipment interchangeable with that of IBM. Traditionally these companies filled the gap when IBM was not able to meet the demand for its products, as is now the case with its 3081 computer, or when IBM introduced new systems not compatible with its own previous models. With United States sales of three billion dollars in 1981, the market share of the PCMs is about 6 percent and rising. Their increasing market share is due, in part, to innovation: PCMs strive to foresee and fulfill market needs overlooked by IBM. PCMs, for example, noted the insufficiency in IBM's disk drives and quickly offered products with double the capacity. But part of their rising market share is due to the obvious nervousness of business as it faces the disarray of the office automation industry: PCMs, like IBM, are a known and trusted quantity. PCM strategy in this area has been to help businesses make the transition to the larger systems needed as central office controllers. The fate of these companies, nevertheless, hinges largely on that of IBM.

The Bunch, an acronym for Burroughs, Sperry-Univac, NCR, Control Data, and Honeywell, produces equipment that is not interchangeable with that of IBM. Although 1981 United States sales were in the thirteen-billion-dollar range, the Bunch's current 33 percent market share represents a decline from previous years. The recent decision of the United States Justice Department to drop its interminable antitrust suit against the Bunch's prime competitor, IBM, does not bode well for these companies. Some Bunch companies are hedging their bets by selling IBM-compatible peripherals. Others are pursuing specialized strategies just to stay alive: Sperry and Burroughs, for example, are attacking markets where IBM has a perceived weakness, notably finance, government, and education. The strength of yet another strategy, diversification, is demonstrated by the experience of Control Data. Fifteen years ago, the firm entered the peripherals and service market as a sideline to its mainframe business, and today it is reaping an abundant harvest. The outlook for the Bunch is not clear;

the companies must correct that dropping market share, and quickly, if they are to have a significant place in the office automation industry.

Mainframes show a 12 to 15 percent unit growth rate, the lowest in the booming information processing industry. Yet the combined expertise of the mainframe makers represents a formidable strength in the office automation battle. In fact, this successfully established industry has the undoubted advantages of more knowledgeable account staffs and wide technical support for their products. And most important, their machines are familiar to the many data processing professionals now running computer departments in almost every large organization. In many cases, these same professionals will be asked to make the decisions involved in office automation. In considering the future of office automation, it is also important to remember that it will be a worldwide phenomenon. Well-established companies like IBM, with its world standard products and internationally experienced sales force, are in a fine position to woo countries just coming into the computer age. And, in fact, such foreign operations are extremely important to IBM. Foreign sales accounted for 45 percent of its revenues in 1982 and 37 percent of its profits. With its mainframe in, say, the central bank, the national university, or a key government department of a developing country, such a company has an almost assured market for smaller computers, terminals, and other computer gear. And when the country begins to search for a model of office integration, who will then be in the perfect position to help?

With growth rates running as high as 45 percent over the past several years, the minicomputer industry is an increasing power in the integrated office race. Buying a minicomputer is often the only way for many intermediate-sized firms to get in on office-of-the-future technology yet still keep costs within manageable limits. The major players in this industry are, in order of importance, Digital Equipment Corporation (DEC), the technically formidable Hewlett-Packard, and the much smaller Data General. It is also significant that the Bunch's Honeywell now does 85 percent of its business in medium and large range computers, but corporate rankings in the minicomputer industry may soon change. In the past ten years DEC has increased its sales by 36 percent to 4.27 billion in 1983. Data General is closing fast, however, for it beat

DEC in bringing out a superminicomputer. DEC's problems stem from poor management and a too cautious marketing approach. To maintain its lead in the face of steep competition DEC must expand into new markets, particularly office automation and personal computers.

The business-oriented microcomputer market should be taking in about $6.3 billion a year in sales by 1987. Although this is small potatoes by information industry standards, recent advances in hardware and microprocessor technology promise much for the future. A few years ago, the term microcomputer meant a small personal machine with limited power and data storage capacity. Today, multiterminal machines with a capacity until recently found only in middle-range minicomputers are meeting the demands of business computing. Such companies as Intel, Motorola, and the Zilog unit of Exxon are in the forefront of technological advance, for they produce more powerful microprocessors. Software, as ever, remains the single greatest impediment to further advance. How important microcomputers will be in the office of the future is thus still open to question.

In general, the markets for all types of computers will remain strong throughout the 1980's. The Gartner Group, for example, projects that the market for mainframes, which had a 15 percent yearly growth rate in the 1960's and 13 percent in the 1970's, will regain its 15 percent rate in the 1980's. Sales of minicomputers, which jumped as much as 45 percent a year during the 1970's, will fall to a still-respectable 25 percent in the 1980's. Microcomputers, though, a negligible force in the 1970's, will surge to the forefront in the 1980's with sales at a phenomenal 50 percent growth rate compounded annually throughout the decade. By 1990, micros, both for the home and office, will be a $300 billion industry.

The United States Bureau of Labor Statistics estimates that there will be a yearly increase in jobs in the domestic computer industry of 4 to 5.1 percent during the 1980–90 decade, comparable to the 4.7 percent annual growth rate for the 1970–80 period. The 1.1 million computer-related jobs of 1978 should increase to over 2 million by 1990. Familiarity with the computer is fast becoming a necessity for job seekers in an economy where the number of computer systems in use will increase from the 600,000 of 1978 to an expected 1.6 million in 1983. Electronic office integration underlines this trend.

The implementation of the office of the future will accelerate demand for technically trained personnel in the information processing industry. But manufacturing computers does not require any skills markedly different from those of the technicians already mentioned for other sectors of the telecommunications industry: roboticists, industrial designers, CAD/CAM experts, and all types of engineers, particularly electronic, electrical, mechanical, industrial, and safety engineers for testing. Although the critical lack of engineers has been much ballyhooed, the real shortage is not in graduates fresh out of school but in experienced computer specialists. Concerned companies now participate in cooperative computer science programs such as that of Ontario's University of Waterloo. Graduates who, by alternating work and study terms, already have twenty-four months of on-the-job experience are immediately snapped up by employers. Since the present practice of pirating other firms doesn't really solve the problem, surely in-house training must be the ultimate answer. Other demand areas include computer technicians who install, test, and maintain equipment as well as operators who know how to make it work. With the escalating use of interactive terminals, the demand for keypunch operators will continue its rapid decline.

The importance of the support sector in every automated office industry cannot be overemphasized. The variety of technical alternatives has created confusion in the minds of potential buyers, which can only be remedied by close attention to education and reliable technical support. The greatest strength of many computer manufacturers, such as IBM, DEC, Hewlett-Packard, and other trusted names, lies in their competent marketing and administrative personnel already well established throughout the world. Competition in the office automation industry is remarkably intense. With many companies manufacturing high-quality products, the cutting edge for vendors lies in constantly improving customer information, service, and maintenance networks. Demand in the automated office industry for those skilled in

- general administration
- finance
- marketing

- promotion
- personnel

will remain exceptionally strong throughout the 1980's and 1990's.

Marketing is, in fact, fast becoming the center of the office automation battle. Although traditionally mini and micro manufacturers leave marketing their product to independent dealers, now even the largest multinationals, such as IBM and Xerox among others, consider it necessary to open retail centers. It is a measure of how seriously competitive companies consider consumer education that they have taken this route of direct access to the public. Retail staffers must be versatile, combining technical expertise with excellent communication skills. Job hunters with these qualities will have no trouble finding employment.

Retail centers also sell software. As we have seen, the lack of quality software is an ongoing difficulty in the computer world. As a result, sales of meticulously crafted software are booming. According to a study of the Norwalk, Connecticut, firm International Resource Development, in 1972 United States software sales amounted to $25 million, but ten years later, in 1982, they were on the order of $2.5 billion, a growth rate of better than 30 percent a year. The figures for Canada are equally impressive. According to the Evans Research Group of Toronto, the 1976 software revenues of $90 million showed a 28 percent growth rate by reaching $253 million in 1980 and should hit $2 billion by 1990. An industry with such outstanding sales growth generates scores of jobs; many of their titles are mentioned in the software section of Chapter Four.

Companies with mainframes usually have in-house data processing departments meeting most software needs, though sometimes independent houses supply systems software, along with plenty of backup support. There are jobs in data processing for programmers, programmer analysts, systems analysts, EDP troubleshooters, and the like, as already mentioned in discussions of other aspects of the telecommunications industry.

For every twenty companies that will eventually implement office-of-the-future technology, only one has a computer today, making the market for all types of software appear almost limitless. Software supply remains a most intriguing possibility for the qualified entrepreneur. Throughout North America more than 5,000 suppliers sell or lease software. The typical company employs fewer than forty people and has sales of between three and five million dollars a year; sometimes the firm has sidelines in hardware or other computer-related merchandise.

In addition, a new expert has come into being—the "software searcher." Many large corporations are setting up whole departments whose primary purpose is to select software. Travelers Insurance Company, for example, which recently ordered 10,000 personal computers from IBM, has its new division, the Information Center, stocked with software and 750 microcomputers. Employees borrow programs from the center in order to appraise the worth of the programs. Eventually the company plans to streamline the system by distributing two thousand micros to agents. Currently there are about fifteen entrepreneurial software search services in operation throughout the United States, ranging from established publisher Elsevier Science Publishing Company to newcomer Redgate Publishing Company of Vero Beach, Florida. Minicomputer data processing managers also generally search for existing software packages rather than trying to produce in-house. Minicomputers use software specific to each individual manufacturer. Some firms, such as Digital Research and Microsoft, license existing software for resale or adapt operating procedures in order to transfer software from one brand of machine to another. Many opportunities exist for firms willing to specialize in this area. Minis also link into larger networks, creating yet another position, that of the network software specialist, who is responsible for designing and implementing the system.

Although current estimates place the market for micro software geared to the business and scientific communities at under $300 million, 1985 should see the figure leap to $2.5 billion and perhaps even $25 billion by the early 1990's. Software needs for business micros are constantly changing. Hardware plays a decreasing role, as more complex software gains prominence. Technically micros for business use have progressed to the point of incorporating complex software systems previously found only in larger computers. Relational data base management systems, for example, once the ultimate in data organizing and manipulating, are now available from some micro software suppliers. Meeting software demands is one of the great challenges of the coming decade. Opportunities for software entrepreneurs remain abundant, but as software production requires little start-up capital, the field is crowded with competitors. Established software entrepreneurs agree that the single most important ingredient for success is aggressive and informative marketing which supplies access to national and international distribution networks. The ideal company displays a good mix of people: some with excellent technical skills, some with ability in marketing, some with experience in management.

LOCAL AREA NETWORKS. The local area network (LAN), sometimes called a data bus, developed by office equipment suppliers, is a relatively recent entry in the office automation industry. It provides an inexpensive solution to the office integration maze: a LAN is a high-capacity link for computers, intelligent terminals, and standalone machines within a relatively small area the size of a building or a campus. Although touted by companies like Xerox, Wang, and Datapoint, LANs are not without problems.

According to the Yankee Group, a Cambridge, Massachusetts, research firm, 2,000 networks worth $15 million were in use in 1982; by 1986, 73,000 networks will engender a $330 million market. While these figures may seem small, the real money is in selling equipment for use on the networks. Wiring an average-sized building, for example, costs from $30,000 to $50,000, but the installation should generate an additional $200,000 in equipment sales. And therein lies a difficulty: most manufacturers sell LANs compatible only with their own office equipment. To some extent the difficulty is technical, resulting from complicated protocols by which machines transmit data. But cynics are surely correct in seeing a calculated attempt by suppliers to keep customers on a permanent, indeed eternal, basis. In an industry where corporate sudden death is very common, the danger is that a purchaser may be left with no source of either service or supply.

The Ethernet System, a joint undertaking of DEC, Xerox, and Intel, is the only network today advertised as compatible with equipment other than its manufacturer's; many manufacturers have begun producing Ethernet-compatible machines. Other players in the industry are Wang, whose broadband cable network is a sophisticated version of Ethernet; and Datapoint, whose Arcnet offers perhaps the greatest flexibility of all, for it incorporates a super PBX. Harris Corporation, a WP vendor, has announced plans to spend $150 to $175 million developing a system to amalgamate all its office communication products. In mid-1983, L.M. Ericsson, a Swedish telecommunications company with $2.7 billion in sales mostly outside the United States, also entered the race. Financially backed by Atlantic Richfield Company (Arco), the seventh-largest American oil producer, with 1982 revenues of $26 billion, Ericsson has a good chance in this area. Surely others will follow.

More technical problems face the industry, however, for the telecommunications conduit to be used wthin a building has not yet been standardized; candidates are two-, three-, and four-wire twisted cable, fiber optics cable, and coaxial cable. Robert Noyce, vice-chairman of Intel, likens the situation to the early days of the railroad industry. Before railroads could provide international freight and passenger service, everyone had first to agree on the width of the tracks. At the moment, those responsible for wiring new office buildings are reluctant to commit themselves and risk obsolescence within a few years.

Another industry issue concerns gateways. The present cost of interfacing LANs with outside telecommunications networks is about $5,000. Intel is reportedly developing gateway chips which could slash the price to a few hundred dollars within the next few years. Most telecommunication is still ground-based, through copper wires. As office automation catches on, the copper tangle will become exactly like a highway during rush hour, causing frustrating delays, the very thing the technology was supposed to prevent. To escape the tyranny of wires, common carriers have taken to airborne transmission via microwave and satellite. Satellite Business Systems (SBS), in a joint venture with IBM, is working on cellular radio for short-distance, intra-urban links between SBS earth stations and user premises. Other companies are working on laser-driven, line-of-sight transmission carriers. This market is worth billions of dollars.

Although the LAN industry is in turmoil today, it must expect growing pains. There is little doubt that LANs will not only capture a significant share of the enormous office automation market but will also, along with companion industries, interconnect and the public telephone system, generate many new jobs. Among them will be, of course, scores of technical occupations as well as support positions. For emerging industries such as these, marketing and public education are vital to future success.

PBX. The modern private branch exchange (PBX) is far more than an enhanced telephone switch. An advanced PBX can act as the central controller in an electronically integrated office. In fact, for many industry observers, the battle for the model of the office of the future has already been won by the PBX. To understand how the familiar telephone came to occupy so august a position, we must examine the tremendous effects of deregulation.

Fourteen years after the Carterfone decision deregulated the phone system, private interconnect accounts for about 37 percent of the installed PBX base in the United States, according to Francis McInerney, president of Northern Business Information (NBI) and publisher of *The Telecom Market Letter* of Toronto and New York. Equipment bought from regulated telcos, especially AT&T, accounts for the rest. In Canada, a mere three years after partial deregulation, private interconnect has already garnered 44 percent of the PBX market.

How Canadian private interconnect could have grown so quickly is an intriguing question. The answer may be timing. Canadian deregulation coincided with the advent of the most advanced PBXs and with increased interest in office automation; the versatile advanced PBXs are a kind of two-for-one sale. Telecommunications managers are under tremendous pressure to reduce telco costs and improve service by replacing obsolete systems. Reduced costs and improved service are interconnect's major selling points. *The Telecom Market Letter,* and insider's dopesheet for the telecommunications industry, provides a convenient example. Brock University, a small school in St. Catherines, Ontario, paid $400,000 for a switch with a life expectancy of ten years plus. Bell rental would have cost $100,000 a year. The new switch incorporates twelve features not available from Bell or only for an additional fee above the rental. By purchasing a sophisticated digital PBX, the telecommunications manager gets not only a modern electronic phone system, but for the same price also gets a device that allows the company to incorporate true office-of-the-future technology. Every institution on the continent is eying that bargain. AT&T has run into some heavy competition.

Selling PBXs has become a very big business, very quickly. In 1982 the global American PBX market amounted to $2.7 billion with private interconnect representing $1.7 billion of that total; industry revenues are expected to reach $5 billion by 1990. The Canadian PBX market was worth about two hundred and five million Canadian dollars in 1982. PBX shipments have been increasing in the United States by about 20 percent annually for the past five years; digital PBX systems have been posting increments of 49 percent a year. Twenty-five thousand digital PBX systems now replace old technology in the United States.

These extremely healthy figures belie the current situation, however, for the industry is about to enter a shakedown period. According to Francis McInerney of NBI, the 1983–1984 growth rate for the industry will be 7 percent maximum and may even be negative. During the past few years, even large firms have folded, leaving the field to twenty-two suppliers, though only five are majors. In 1982, the leading manufacturers of telephone equipment were first, Western Electric, a subsidiary of AT&T; tied for second place, two Canadian companies, Mitel and Northern Telecom, 55 percent owned by Bell Canada; and in third place, Rolm. The former Terryphone, a subsidiary of International Telephone and Telegraph (ITT), now called Business Communications Systems, must also be mentioned. ITT's strategy is to transform its subsidiary into the largest interconnect company on the continent, a true rival for AT&T.

The interconnect industry faces difficulties in the short term for several reasons. The enormous number of shipments already made comes close to saturating the present market; competition is also intensifying within the industry. In 1984, when AT&T divestiture has been completed, the regulated equipment industry will no longer exist. The seven new regional telephone companies, which will operate local service, will no longer be required to buy equipment exclusively from AT&T's Western Electric subsidiary. Since Carterfone, AT&T has protected itself with what can only be called a defensive marketing strategy; it has merely replaced old systems within its already established customer base. Meanwhile, other private companies seized the initiative in the new or growth markets by quickly developing competitive state-of-the-art products. By 1982, however, AT&T had run into a solid wall of market resistance. It had already replaced about 75 percent of its customers' old equipment; the rest looked to the more innovative companies for product. To add to its woes, AT&T faced massive reorganizational difficulties brought about by its agreement with the Justice Department. Mitel, by this time, had swamped AT&T in the small system end of the market, while Northern Telecom had gobbled up the large system end. AT&T, eaten away at both ends, is now left defending the middle market.

AT&T must also now prepare to do battle with IBM. In assessing the future of this market, it is worth remarking that IBM, which already owns 12 percent of Intel, a leading semiconductor producer, bought 15 percent of Rolm, a Santa Clara, California, PBX manufacturer with 1982 sales of $381 million. IBM's announced goal was to increase access to its computers through Rolm's equipment which now controls 49 percent of the digital switching market. As IBM competes with AT&T, however, so does AT&T compete with IBM. After divestiture, AT&T will be able to use Bell Labs and its Western Electric subsidiary to compete directly with IBM. Through the new American Bell, AT&T is expected to introduce a line of computers in 1984. In addition, System 85, its super PBX, and the Antelope, still under development, show that AT&T is far from beaten.

TRANSMISSION. The PBX and LAN models for the office of the future bring us to the brink of another enormous billions-of-dollars-a-year industry, transmission outside the office. It is here that the boundaries separating the various office automation models begin to blur and merge into one another. Although space does not permit an extended explanation, no discussion of the office of the future would be complete unless it gave at least some small inkling of the complexities of the transmission issue.

The telephone system, which handles all public voice transmission, is regulated by the FCC, while data transmission is not. IBM, for example, may not offer voice transmission services; telephone companies may not offer data transmission services. In fact, however, the phone network can now store and analyze information (data) as well as transmit it; computers can "talk" to each other and can do so over telephone lines. The information, or data processing industry, and the communications, or voice transmitting industry, clearly separated when the regulatory laws came into force, have now converged, due to their common use of digital technology. Companies in the voice transmission business and those in data transmission have devised means, many of them ingenious applications of state-of-the-art technology and all of them perfectly legal, to circumvent the obsolete FCC regulations.

The domestic communications industry is now worth $50 billion annually; data transmission is a $22 billion-a-year industry. Both industries display good, healthy 15-percent-a-year growth rates. Archaic laws now confine companies to assigned turf, but all the players are intensely eager to jump the fence and get into forbidden territory. Automated office technology gives them the chance.

To this end, AT&T proposed its alternate model of the automated office, Advanced Communications Systems (ACS), which was to have enhanced the intelligence of the phone line to the point where it can completely integrate all transmission, voice and data, both inside and outside the office. Though not mentioned much recently, ACS was to have been a direct threat to IBM's model, Systems Network Architecture, which establishes standards best suited to IBM equipment. Of course, even though AT&T with its $115 billion in assets and its near-monopoly control of the entire American telephone system would appear to be sitting in the catbird seat, this largest company in the world clearly fears the coming battle. And with good reason, for IBM, LAN manufacturers, and many others have figured out a way to grab some of Bell's best business.

Research shows that up to 80 percent of a large company's communication flow goes merely to different branch locations of the same firm. Intracompany communication, now accomplished mainly by telephone, is a tremendous market for AT&T: the twenty-five largest telco users in the United States spend $1.2 billion a year, doing 15 percent of all long-distance calling. The entire intracompany business communication market is estimated to be worth $7 billion a year. The top thousand United States companies, such as General Motors, Ford, and General Electric, account for 65 percent of all intercity phone business. Competitors are waiting to pounce on that 65 percent, now a regulated Bell monopoly.

IBM, for example, in a joint venture with Comsat General and Aetna Life and Casualty, offers its Satellite Business Systems (SBS) as a way for a client company to meet all its internal transmission needs. Since the Carterfone decision, internal voice transmission is not regulated; it is perfectly legal to wrest this business from Bell, although, up to now, interconnect generally handled only calls within a building. Now IBM aims to poach some of Bell's long-distance business: the customer buys an earth station for each company location and then pays SBS a fee for the privilege of bouncing all internal voice and data transmission off its satellite. For up to 80 percent of the SBS client company's communication, then, Bell is completely out of the picture. IBM gets to keep all its data transmission business, plus a nice chunk of Bell's $7-billion-a-year market. LANs obviously could employ a similar long-distance communication method. RCA is also offering satellite data transmission. Bell is fighting back through Northern Telecom, which is 55 percent owned by Bell Canada and the second largest manufacturer of telecommunications equipment in North America after Western Electric, a subsidiary of AT&T. At the end of 1982, Northern Telecom announced a $1.2 billion five-year development program to produce goods and services for its Open Protocol Enhanced Network, known as OPEN World. OPEN World allows all voice, data, and graphics transmissions from a variety of systems, including IBM's Systems Network Architecture, to be routed through the Bell central switching office. Thus, all inter- and intra-office and inter- and intra-company communications can take place through Bell facilities and equipment.

For the job seeker, all this means that there is and will be plenty of action. In their duel of the titans, the corporate giants will need the best hired hands they can get. That means both those with technical skills and those with support skills will find an outlet for their talents. The office automation industry is just getting under way. The best time to get in on this hot new market is now.

SUBSIDIARY MANUFACTURE

Primary equipment manufacture is a game for only the biggest players, generally the largest multinationals in the world with revenues in the billions. While the office automation industry would thus seem to leave nothing for the individualist who wants to go it alone, nothing could be further from reality. Entrepreneurs are an intrepid lot; they know that even gigantic corporations cannot possibly cover every aspect of this huge market. Entrepreneurs search out the lucrative market niches overlooked by the giant companies. Although office automation is still in its earliest stages, many have already become multimillionaires using this simple method. The vital and growing industry provides the opportunities; it is up to the alert, the clever, and the bold to take advantage of them.

Adventurous entrepreneurs are, in fact, as essential to North America's continued economic well-being today as they have always been; entrepreneurs are our greatest asset. In the competitive world of office automation, the Europeans and the Japanese are ready to go for the jugular at the slightest sign of weakness. Lacking a native tradition of hardy entrepreneurship, however, the Japanese and the Europeans also lack the market flexibility of the Americans. Entrepreneurs are not bound by company policy or lines of corporate command. Entrepreneurs are free to act at a moment's notice and to follow their own best instincts—their business indeed depends on it.

Jesse Aweida is a fifty-year-old Palestinian refugee manufacturing computer memories in Louisville, Colorado. His company, Storage Technology, did $64.7 million worth of business in 1982. American responsiveness to drive and ambition allows that to happen. Sirjang Lal Tandon, another immigrant, manufactures the disk drives Radio Shack puts in its personal computers. Founded in 1975, his firm, Tandon Corporation of Chatsworth, California, had 1981 sales of $54.2 million. Phil Hwang came to the United States from Korea. In 1975, with $9,000 in family savings, he started his company, TeleVideo, to produce computer screens and keyboards. Today, Phil Hwang is a multimillionaire. The spirit of adventure and the willingness to risk all for an idea are characteristics of successful entrepreneurs. These traits have been the backbone of America's glories in the past and will continue to be so in the future. Immigrants with skill, energy, and ambition can make it on this continent because the North American economy has always encouraged and rewarded these qualities.

When times are tough, North Americans respond with renewed energy and vigor; the entrepreneurial spirit surges to the fore, just as it is doing today. In 1981 some 587,000 companies filed incorporation papers, an 80 percent jump over 1975 and 53,000 more than in the previous year. Of course, many new businesses fail, but high-tech enterprises related in some way to computers or office-of-the-future technology are the most likely to succeed. The vast majority of Americans work for large corporations. Yet during the past ten years, the largest one thousand firms in the country have done little but maintain a constant number of employees. Small businesses, on the other hand, generated over three million new jobs during the same period.

Working for a large corporation has its advantages, as generations of Americans can testify: a regular paycheck, paid vacations, job security, health insurance, a pension plan, and company benefits, possibly including stock options, a car, mortgage assistance, paid moving expenses, and more. The entrepreneur sacrifices this security for long hours, an erratic salary, no vacation, no insurance, and a chance to make more money in a year than during a lifetime of nine-to-fiveing.

Often entrepreneurs learn their trade and hone their skills in a large corporation before setting out on their own. Gordon Matthews worked for IBM and Texas Instruments before taking the plunge. His third entrepreneurial endeavor, ECS Telecommunications, which sells a computerized system for storage and transferring messages by telephone, did four million dollars worth of business in 1980, its third year of operation. Even working for an enterprising small business person, while not as secure or as well paid as corporate life, is a good way for the would-be entrepreneur to get started. In a small company, employees can learn the entire business.

Market niches overlooked by the big companies in office automation are not hard to find. Proximity Devices of Fort Lauderdale, Florida, founded by a twenty-five-year-old mathematician and grad school dropout, markets a spelling error detector and 89,000-

word English lexicon for computers. The company hit it big when IBM decided to offer its product as EasySpeller, a $125 option on its personal computer; other micro and WP manufacturers will soon follow suit. Other ventures are still in the developmental stage. Cognex Corporation of Boston, for instance, is marketing a sophisticated optical character recognition system, called DataMan. Launched in 1981 by five investors with less than $250,000, the company has already sold to IBM and Kodak; it had sales of $1 million in 1982. Integrated Market Systems in Silicon Valley is working on a turn-key work station to service all the office and road needs of sales organizations; first deliveries take place in 1983. Its founder, John Peters, predicts $100 million in sales by 1988. Pipe dreams? Maybe, but don't bet on it.

Destek, Inc., fills another market niche. Office automation multinationals, occupied with bigger fish, have ignored the needs of small businesses with a mere five or ten micros. Destek manufactures a low-speed LAN which can interconnect a small number of micros. Founded in 1980, the firm had 1982 sales estimated at about five million dollars. Another group of business people have spotted a momentary weakness in LAN marketing: their network bridge concept allows computers to communicate with each other and with computers in other networks. The companies do not sell complete LANs but connections or "bridges" to them. Intel's gateway chip would make their product obsolete. No matter, companies like Interlan and 3Com are ready to change their turf at a moment's notice. In the meantime, they could make a killing in this market with plenty of time to plan their next venture. Concord Data Systems of Lexington, Massachusetts, is trying to produce its own LAN. During the long developmental stage, the company is selling modems. Both its founders come from Codex Corporation, a Motorola subsidiary and market leader in high-performance modems, so the market was a natural. Flexibility is the password.

Innovative office automation products can actually create entire subindustries, worth billions, almost overnight. The introduction of the rigid disk is a good example. A mainframe's memory is internal, stored on dishwasher-sized disk drives holding up to 500 million characters of memory each. Memory storage in a micro, generally located externally in its disk-operating system, is both clumsy and limiting. The Winchester rigid disk aims to remedy the situation. An internal memory which cannot be removed from the micro, the Winchester measures only 5½ inches in diameter; it can hold up to 10 million characters of memory, the equivalent of three large novels. The Winchester's greatest drawback is its price, $3,000 to $5,000, compared to about $700 for a floppy. The Winchester market currently runs about $81 million annually, but the California research firm Roman Associates International foresees growth to $1 or $2 billion a year within a short time. Eventually the Winchester will account for one third of the rigid disk market, which in 1980 totalled $3.68 billion. In 1981, for example, only one company, Seagate Technology, was working on the Winchester rigid disk; by 1983 there were more than ninety manufacturers of rigid disks in Silicon Valley. In this booming field, however, you never can tell. Current developmental projects include a cartridge rigid disk, multiple floppy disk cartridges, and from the Japanese, reports of a 3½-inch rigid disk.

Even the seemingly monolithic telephone business supplies opportunities if you know where to look. In fact, the telecommunications industry is moving so fast today that few realize just how extensive it has become. Alert entrepreneurs are now offering business customers long-distance telephone call service at 20 to 60 percent discounts. Miami's Teltec Saving Communications Company, for example, leases dedicated lines from Southern Bell and others at a fixed price. Then the firm packs the lines with round-the-clock discounted client calls using a computer timesharing system. Begun in 1980 with 9,500 long-distance calls a month, the firm now handles 450,000 calls a month. Telecommunications is an entrée business, offering unlimited possibilities: starting with one type of signal, usually voice, a company can expand indefinitely to data, video, facsimile, and electronic mail.

The accessories market is another good place for entrepreneurs of all kinds. The big companies are not able to service this aspect of the office automation industry; it is up to the intelligent observer to find a need and fill it. Such a firm is Datamarc, Inc., a Dallas-based firm specializing in the manufacture of paper-handling equipment attaching to WPs. The firm sells an automatic envelope feeder to complement the five most popular WP printers on the market, a form feeder, and a sheet feeder for smaller offices.

In addition, companies manufacturing floppy disk holders or any of a variety of small yet convenient office necessities can count on a steady market. New machines create niches for gadgets and gizmos.

New machines also expand the market for the office furniture which holds all this equipment both esthetically and conveniently. Traditionally the furniture industry suffers in a recession, yet office furniture has been one of the fastest growing industries in the United States in recent years. Sales of office furniture, amounting to about one billion dollars in 1975, had quadrupled to around four billion by 1982. Systems furniture, used in open-plan offices, was first introduced in 1968 to help integrate computers into offices; it now accounts for 32 percent of the market. The proportion of workers in open-plan offices increased from 30 to 36 percent in just the two years from 1978 to 1980. Some experts say that within six years 60 percent of the work force will be using computer terminals. Sales of specialized office furniture will surely continue to mount. In fact, research indicates that office furniture may be more important than most people realize. A survey by Louis Harris and Associates for Steelcase, Inc., an office furniture manufacturer, reveals that many workers would even accept lower pay for the chance to control heating and air conditioning, for more privacy, or for better back support in their chairs. Designers and manufacturers evidently have a large potential market here.

The increasing tangle of wires necessitated by office machines is another challenge to furniture manufacturers. Some desks now have hidden grooves underneath to accommodate cords. Wires are proliferating so rapidly, however, that furniture manufacturers can only provide a small part of the solution. Buildings themselves need a new look. Anyone who has ever visited a large new computer installation in an older building has seen the problem cords present: sometimes a false floor is installed to keep the unsightly scramble from causing accidents. Architects could surely come up with better designs while renovating older buildings. New designs will have to take the wires into consideration. Architects should have plenty to do.

Computers and their accompanying machines need an entirely different architecture. Interior designers and decorators are even now specializing in "environments" for computers and computer personnel. Sometimes computers require special buildings, which means that architects have another outlet for their abilities. In addition, heat is a problem wherever there are computers. Those specializing in fire prevention and thermal control must contribute to the architect's final model of a computer installation.

With so much going on, the automated office industry certainly appears chaotic. But the chaos is fertile. Probably no other industry provides better opportunities for carving out jobs and creating employment. Some people's problems, after all, are other people's jobs.

CHAPTER EIGHT

LOOKING BEYOND THE OFFICE OF THE FUTURE

LIVING WITH AUTOMATION: LEARNING TO LOVE CHANGE

The electronically integrated office is a technology in the very early stages of implementation. Its possibilities are multitudinous, but its future is by no means certain. Its capabilities have in no wise been fully revealed or tested. Throughout the 1980's and 1990's, equipment manufacturers and corporate decision makers alike plan to introduce the automated office on a vast scale.

Considering that more than half the American labor force works in some kind of office, we begin to see how immense will be the effects wrought by implementing new technology there. The lives of millions of working people, from the lowliest clerical to the loftiest chairman of the board, will change as their workplace alters forever. Change is never easy for human beings; change on so vast a scale is bound to cause substantial trauma to many, many people: those already employed and even those about to enter the labor force. As working conditions are transformed, routine and habits must alter. It is no wonder that many in the work force anticipate the automated office with fear. They are right to be afraid. Already in its short history, office automation has caused horrific difficulties along with its astronomical gains in productivity. How employers view the purpose of technology is one important aspect; how employees view it is another. Past experience has not been reassuring.

For some employers, the whole point of automating the office is to save money by eliminating jobs. Metallics, Inc., a Bristol, Connecticut, manufacturer, saw office automation as the only way to avoid bankruptcy. Since computerizing the office, the company has let go fourteen out of twenty-one clerical workers. A recent study conducted for the Air Force Systems Command, the purchasing arm of the Air Force, found that fully automating its offices would have the same effect as increasing its clerical staff by 25 percent. Although office automation would cost twenty-five million dollars over a five-year period, such an outlay would still be cheaper than hiring more people to do the same job. European studies predict a 30 percent loss of jobs in the French banking and insurance industries and big layoffs in the British clerical work force as a result of automation. When office-of-the-future technology is implemented for the purpose of reducing staff, employee anxiety is, of course, a foregone conclusion.

Closely allied to the concept of using automation to eliminate jobs is another heavy-handed idea: re-creating a factory assembly line in the office. The first wave of word processors, for example, replaced typewriters in secretarial pools; operators could then spend the whole day doing nothing but typing. Some companies isolated operators in computerized "data centers" where pay was low and opportunities few. WPs also allowed the employer to establish precise timetables for certain types of work, such as 3.6 minutes to process document A or 4.3 minutes to process document B, and to identify workers failing to meet the standards. Other productivity quotas, such as a minimum number of keystrokes per hour (in one case, as many as 21,000), were also instituted. In short, the office of the future became the factory of the past, causing the same worker alienation, absenteeism, and stress.

Although stress, in the popular conception, is part and parcel of the jobs of brain surgeons and high-ranking executives, in actuality those with the most boring jobs and standing lowest in the pecking order are the most likely to exhibit the classic symptoms of stress. Numerous studies have shown over and over again that stress results from impersonal working situations and loss of control over the environment. According to a study made by the National Institute for Occupational Safety and Health, secretaries rank second as the victims of stress-related diseases. According to the Canadian Institute of Stress, the country loses thirteen billion dollars a year in lowered productivity, lost work days, disability payments, medical payments, and replacement personnel as a result of stress-related disabilities. Applying this figure to the United States, which has ten times the population of Canada, would yield losses on the order of $130 billion a year. Obviously, implementing the office of the future is hardly worthwhile if it causes that total to rise.

Computerizing the workplace without careful thought or consideration for the employee can have other terrible and unanticipated consequences. Computerized cash registers can keep track of how many transactions per hour a cashier handles. In Denmark such electronic surveillance has already caused operators to walk off the job. Computerized access-control systems, found today in any number of large corporations, require employees to punch in with a machine-readable badge for security purposes. Incidentally the badges leave a computerized trail, pinpointing a worker's whereabouts at all times. When a French company took the process one step further and installed the system on restroom doors, the employees went out on strike. Privacy can become a thing of the past in the integrated office. In electronic mail systems the employer customarily receives an accounting of who sent and received messages as well as long-distance calls. Employers can also examine the contents of an electronic system. In 1979, some employees at Sandia National Laboratories in Albuquerque were discovered using office computers for a bookmaking operation. An electronic "sweep" of the whole system showed workers storing everything from bowling scores to personal letters. In the traditional office, for management to inspect desk drawers would be an invitation to open warfare. Although an electronic sweep would seem to be akin to rifling drawers, employers apparently did not consider it an invasion of personal privacy. Such administrative sangfroid raises dim forebodings about the future of us all in our brave new world.

As the horror stories multiply, it is clear that lower-echelon workers have very good reason to fear office automation. And now it is the turn of managers to start quaking in their boots. The business world has recently awakened to the fact that nonclericals, whose salaries represent 66 percent of all office labor costs, have demonstrated the lowest productivity gains. Their work, generally unstructured or semistructured, contributes mightily to the escalating costs of running offices. So the focus of office automation has shifted to the work of managers and professionals. Today, the objective in automating the office is less to facilitate structured work like typing and more to improve the effectiveness of managers, administrators, and executives. Various studies have shown that professionals spend 15 to 40 percent of their time on "subprofessional" tasks. Putting numbers together for reports, copying, and scheduling could be assigned to secretaries or computers, leaving highly paid managers free for more creative and productive activities. Office systems are now capable of facilitating communication among nonclerical employees, as well as supporting managerial decisions, by providing reams of relevant, organized, and topical information. Office automation can provide tools for unstructured decision making, planning, forecasting, and other management functions.

The organization of the modern office is hierarchical in the extreme. It ultimately takes its form from the medieval Church and the European army of the sixteenth century. Implementing the office of the future will affect, if not forever transform, every single management position and function. In the electronically integrated office, gigantic communication networks connect everyone. Machines do not distinguish between clerks and vice-presidents. All have equal access to the system. Providing instantaneous access to information and supplying the enormous computing power of the most advanced systems to everyone will have revolutionary consequences. The current model for management is a ladder; each rung represents a step up to greater power and responsibility within a company. The ladder model will flatten into the circle or wheel paradigm, for on electronic networks all workers have equal access to information and computing power. Such a democratization will have significant and, as yet, not well-understood consequences. One thing, though, is clear: as the organizational model shifts, so does the power. Accustomed to one well-understood road to success, executives and managers are frightened of losing their secure position within the hierarchy. Their worldview must change and, along with it, familiar ways of doing things. Having seen the implementation of automated office technology bungled at the clerical level, thoughtful professionals have every reason to fear the worst.

Implementing the office of the future could be, in short, a nightmare for clericals and executives alike. Fortunately for millions of office workers, however, decision makers are learning from past mistakes. Undoubtedly electronically integrating the office results in hard dollar savings: an automated office uses less expensive floor space, more productive employees put in far less overtime, the need for costly temporary clericals is reduced. Yet, as business is discovering, apparent hard dollar savings can disappear as if by magic. Initially, when the office of the future meant improving clerical efficiency with word processors, administrators, lacking foresight, simply replaced typing pools with WP centers. But isolating operators in WP ghettoes led to unnecessary stress, alienation, and as a consequence, sometimes downright sabotage. Gradually managers became aware of the extraordinarily high hidden costs of ham-handed office integration. Bored workers get sick and cost the company money. High turnover rates mean lost training time. Bored workers walk out the door, taking along skills learned on the job at company expense. In retrospect, it is easy to see that saving a few dollars by creating office assembly lines is hardly worth the effort. In the long run, high productivity gains from powerful technologies are merely cosmetic if the human concerns of employees are ignored. Careful planning is a necessity if the full capabilities of the office of the future are to be realized. As a small first step, in the more advanced automated offices, secretaries are now returning to their former places.

The soft dollar savings made by improving management effectiveness are now the focal point of any serious implementation of automated office technology. And as the technology evolves, office systems geared to the needs of managers are appearing with increasing frequency. Electronic mail, for instance, could save an executive as much as several hours a day. Such soft dollar savings, however, are not only difficult to measure and therefore cost-justify, they also raise other complex issues. What does the manager do with this new time? The key to productivity gains lies in developing a viable reinvestment strategy. Gaining precious time for high-priced management staff means little if the time is not used creatively.

At this point, automating the office may appear as a damned-if-you-do, damned-if-you-don't situation. On the one side, business, staring at the specter of aggressive competitors, low productivity, and vanishing profits, must automate; on the other, worker sabotage and management alienation loom as possible consequences. It would be foolish to consider only the rosy side of office integration, for such conundrums are the very real problems facing corporate decision makers today.

Luckily a whole new sector of the office automation industry has recently come into being for the express purpose of easing the new technology into place with a minimum of upset. As is now obvious, the changeover can achieve optimum results if it proceeds in a carefully organized and planned manner. Orchestrating the activity are what may be called agents of change, a catchall term including people from diverse backgrounds. These change managers must have a foot in both worlds: they must understand technology, yet remain humanists at heart. Ensuring that the horror stories of the past are not repeated, they will make sure that the automated office increases productivity because it improves the quality of working life. Change managers will help design systems which take into account technological, social, and environmental concerns so as to produce the interesting, meaningful, and autonomous work which guarantees higher performance. Change managers are crucial to the future of the automated office. Joining their ranks, which will be legion for quite some time to come, is a new and exciting career option for those with open and inquiring minds.

Change managers, under a variety of titles, work in many situations: in large corporations, already well advanced in automation; in traditional management consulting firms, which are belatedly jumping on the change bandwagon; and in small innovative firms devoted exclusively to change management. Change managers can come from anywhere. As in baseball, where a smart player, coach, or scout can become a field manager, so in this world of office automation, any of a number of people with a variety of occupational titles can become change managers.

The Bank of America Corporation, for example, reports that expanding automation is saving the company millions of dollars by cutting paper work and boosting productivity. Heavily involved in word processing since the mid-1970's, the bank saved seven to ten million dollars in just the first few years and then, according to spokesmen, stopped counting. In fact, the California-based bank finds automation so advantageous that some 140 professionals work full time researching, consulting, and planning further steps. With change managers guiding the operation, bank employees have been adequately prepared and consulted. Now, encouraged to voice their desires, workers themselves contribute many of the most effective cost-cutting applications of the bank's computerized system. The Bank of America's change managers have shown that automation, if wisely implemented, can enhance the jobs of office workers. Standard Oil, also of California, is another company in the forefront of office integration. From word processing to electronic mail, the company's prime purpose in computerizing is to improve the productivity of its professionals. Through automation, professionals can assign their routine administrative tasks to clericals. Clericals, freed from meaningless office drudgery, are more productive. Professionals, under the guidance of change managers, find more creative ways to spend their time. Both groups show incremental productivity gains, and employee morale has taken a sharp upswing.

Recently organizations providing consulting services to all sectors of the office automation industry have realized the enormous importance of the emerging change manager role. On the data processing side of this complex industry, all big management consulting firms now have office automation programs, as do large accounting houses, data processing consulting firms, and systems houses. On the word processing side, change managers can be found in almost all office-products and word-processing consulting firms. On the communications side, most telephone consulting companies, such as the Durham Group, ICA, and Telemanagement, have shown interest. In fact, almost every large corporation in the industry is now getting into the act. New job titles, all of which translate to change manager, are appearing every day: corporate communications director, vice-president for system programming, corporate telecommunications head, manager of office automation, and integrated systems specialist are some of the many on the list.

Small change management firms are also coming into existence. Trigon Systems Group of Toronto, founded in January 1982, is a consulting firm specializing in planning, designing, implementing, and supporting automated office systems. Its three founders, Don Tapscott, Morley Greenberg, and Del Henderson, all formerly with the Office Information Communications System program at Bell Northern Research, are now change managers. Each brings totally different skills to the enterprise: Tapscott's background is in research methodology and data processing, Henderson's in education and marketing, and Greenberg's in industrial psychology. Although the particular skills of change managers may vary considerably, the firm's personnel must be multifunctional in order to address the complex issues raised by the electronically integrated office. A firm such as Trigon plays several roles simultaneously: it serves as a bridge between the vendors and the users of integrated office systems; it develops innovative planning methodologies; it investigates new organizational forms required by the introduction of a new technology.

To fulfill these simultaneous roles, change managers must first of all discover what the marketplace needs; so they conduct studies. While still at Bell Northern Research, Trigon's founders participated in a research project designed to assess the effects of an electronic office system on its users. Another purpose of the study was to experiment with testing and interviewing techniques for collecting and evaluating information from the test group. For this closely monitored group, the study showed that as employees gained improved access to information by using sophisticated office systems, the amount of information they perceived as necessary to do their jobs increased. And as members of the group learned how to apply the new technology to their jobs, they grew more positive about it. Besides demonstrating how enthusiastic people can become about automated office systems under the right conditions, this project and others like it yielded further important results. To some extent, manufacturers are in the dark about the real needs of potential users: the test group had many suggestions about product design and ways to successfully market the equipment. This group's experience paved the way for implementation of the new technology elsewhere.

The Trigon staff conducts studies before, during, and after implementation, all with the full participation of the target group. The before research is a feasibility study assessing costs as well as technical, physical, social, and organizational needs of the individual client company. From this study, Trigon staffers tentatively determine which equipment would be best for a given situation and where it should be placed. Next, a pilot project, affecting only a small segment of the company, is inaugurated. The selected group constantly voices opinions and reactions by keeping diaries, giving interviews, and answering the change managers' questionnaires. Then, if the pilot project so warrants, full implementation proceeds, again under the close guidance of the change managers. The monitoring is an especially important part of the whole process; it often yields surprising items. In one company, for example, no one, it was found, was using the electronic mail function of the equipment. Questioning revealed that people did not know that it existed. A final step in the implementation process includes post-tests which indicate how to refine and extend the system. Post-tests also measure the effectiveness of the implementation and supply information which could ease the pains for another organization.

Aside from bridging the gap between vendors and users, a company like Trigon is suggesting ways of redesigning not only jobs but also the entire hierarchical order of a client company. Very few people understand just how pervasive the changes in the new office will be; they will leave virtually not one job untouched. To give just one small example, personnel positions will radically alter. Personnel managers must fully understand the potential effects of the new technology in order to prepare contracts, decide on working hours, and enter into union agreements reflecting changed working conditions. Even recruitment will entirely change.

Already the organizational structure of offices has begun to change as executives turn to popular studies of Japanese corporate life. They think they are boning up on the tactics of the competition, the infamous Japan, Inc. We know better: such an interest in an alternate managerial model is the first stirring of the inevitable philosophical revolution that is the direct consequence of the introduction of a new technology. We are always fundamentally influenced by the tools we use. The introduction of cheap automobiles like the model A or the model T contributed to the development of our modern nomadic urban existence, exemplified in the drive-in movie, the car pool, suburbs, expressways, and the fast-food joint. So, too, the automated office will initiate an equivalent upheaval in future forms of organization.

Guiding the revolution, change managers will be around for some time to come. For this revolution will take many years to complete. Since change management as an occupation is only now coming into being, it cannot be described in detail or defined with great precision. No one can get a degree in change management nor can one apprentice. We do know the problems which must be addressed, however, and we can speculate about the kinds of skills needed to solve them. The following professionals are among those who might have something to contribute to the implementation of the office of the future:

- computer professionals
- data processing experts
- information scientists
- management scientists
- labor relations specialists
- organizational development experts
- sociologists
- industrial psychologists
- specialists in testing and measuring
- communications professionals
- social psychologists
- research methodologists
- statisticians
- economists
- anthropologists.

Although this list is a long one, it is by no means exhaustive. Communicating clearly and articulately is obviously an important skill for change managers, but above all else, they must have the ability to listen, a most vastly underrated part of the communication process.

Firms like Trigon and its counterparts in the United States, such as Advanced Office Concepts, Ltd., and Office of the Future, Inc., or consultants to the vendor side, such as the Yankee Group and Diebold Office Automation Program, fill an enormous need. Change managers do a lot more than act as go-betweens for equipment makers and offices. Change managers are philosophers, psychologists, cheerleaders and coaches, conductors and choreographers. They are teachers. A change agent must be multifunctional and multidisciplinary, combining computing, communication, and management skills. The ideal change manager is so multifaceted a character, in fact, that such people do not really exist. The ideal change manager is the Renaissance humanist, the fabled man for all seasons, come back to life in the twentieth century. Failing that, real change managers are usually those who can combine a strength in one technical area with an overall comprehension of the potential impact of the coming revolution. For firms like Trigon, the only real constraint to unchecked growth is lack of personnel.

Managing the revolution will require efforts on all fronts, but no effort will be more extensive than in the education sector. To fulfill its role of supporting and interpreting the new technology, the education sector takes a number of unexpected forms, all of them providing ample entryways into the thick of things. Colleges and universities are, of course, the traditional educators for business as well as most other sectors of the economy. Today these institutions are making stringent efforts to keep abreast of new developments occurring as a result of the transformations wrought by the new technologies. Junior colleges, colleges, and universities offer night courses, weekend seminars, skill-building courses, and a variety of other programs.

Other, less well-known educational institutions exist, however, and are growing at a rapid rate. The American Management Association (AMA), for instance, along with its affiliate, the Canadian Management Center (CMC), is a nonprofit organization, founded in 1923, which attempts to keep business professionals up to date. The AMA offers over 4,000 short, practical courses a year and the CMC about 200; most of the courses are of one to four days' duration. Working business people teach the courses, usually on a straight expenses or fee-for-service basis, which have titles such as Production Planning and Control, Utilizing Computer-Aided Engineering for Better Design, Preretirement Planning, Career Workshop for Executive Secretaries, and Improving Interviewing Skills. The students, too, are working business people. The advantage for the teachers in such an organization is that, although they make no money for their efforts, or a very minimal amount, they do gain exposure for themselves and their companies. Students hone their skills and keep pace with new developments, of course, but also, by participating in the course with a peer group of those actually in the workplace, they have the opportunity to make new business contacts.

The AMA provides a broad range of services. Recently it began offering courses on integrated office systems, which have proved to be both popular and useful; these courses will help many in making the changeover to the automated office. For those unable to leave their place of business, the AMA provides in-house courses for groups or supplies multimedia packages and a trainer for the companies to use. *Management Review Magazine,* published by the AMA, supports its courses through the written word. In addition, the AMA's Amacom arm is the largest publisher of business books in the world. The AMA also publishes an annual Executive Compensation Survey; it has compiled a computerized data base answering such pertinent questions as: What are the rules and regulations for exporting a certain product to Saudi Arabia? What does a PR person do? Finally, the AMA runs a center for strategic and corporate planning in Hamilton, New York, to help top executives in long-range planning. All these activities support, in one way or another, the coming revolution; they also indicate the endless possibilities offered by this sector to the imaginative job seeker.

Yet another educational area has quietly grown to truly gigantic proportions parallel to the traditional university: the corporate educational system. Assessing the actual size of this parallel system is impossible. Knowing that every large corporation has an educational sector gives some indication of its extent. Charlotte Mallouh, now section manager for systems education operations at Bell Canada's Montréal office, has participated in corporate education almost since the beginning of her career. When her children grew older, Charlotte returned to university to take data processing courses. Throughout most of her university education, she took her courses at night while working during the day at various large computer companies, such as Univac. After gaining valuable teaching skills by demonstrating small systems to managers and technicians, Charlotte joined the faculty of a newly opened private college, where she developed the data processing curriculum. Eventually she joined Bell Canada in order to learn more about the most advanced large computer systems; she knew that Bell always has the best. Beginning as an instructor in computer languages, within a short time Charlotte became section manager of the data processing educational program, called systems education operations.

With its six instructors and three clerks, systems operations is the smallest group in the entire education division. Offering a total of about one hundred courses, most modularized for use with videotapes and a handbook, the section also gives twelve classroom courses, each of two to five days' duration; course enrollment varies from twelve to twenty students. Since Bell constantly upgrades and expands its exceedingly sophisticated computer systems, its data processing and other personnel need the courses in order to keep pace. Courses have titles such as Building a Dictionary for Designer-User Communication or the Methodologies of Program Analysis and Design; there are many computer language courses. By taking these courses, Bell workers are always up to date on the latest technology. But the cost is high. An incredible 15 percent of the time of all data-processing personnel at Bell goes to training, including an average of 5 to 10 percent per year, a full two weeks to one working month, in formal courses such as those in Charlotte's section. Few corporations can match that record for dedication to excellence.

Computer training is only one area of education at Bell; there are courses for controllers, in internal finance, in management training, and much more. Bell even has a sector for general business skills, giving courses in reading, interpersonal communication, effective writing, interviewing, and decision making. Corporations like Bell cannot afford to wait around for skilled personnel to walk in the door. Change at Bell can occur relatively quickly because the company does its own in-house training. Bell has always demonstrated an exceptional interest in the self-and-career-development of its personnel. Indeed, the company constantly requires such development by asking groups and individuals to prepare presentations for others and, in particular, for higher management. In this way, the individual has a chance to learn new skills and at the same time to show himself to best advantage before superiors. Management is, by this method, always aware of activities at lower echelons: this addresses the perennial problem in large organizations, communication among the various levels of authority. At Bell, contrary to the popular adage, the left hand does have a pretty good idea what the right hand is doing.

What Bell does for its employees is not all designed to get them through the changeover to the new world of the automated office, but a surprising amount of it is. People who are paid to spend 15 to 25 percent of their working time learning are really spending

15 to 25 percent of their time changing; for learning always implies change. The emphasis which Bell, a company in the vanguard of the communications revolution, places on corporate education may be exceptional today, though IBM reportedly invested more than $500 million in 1982 on employee education and training, yet very soon it will become the norm. It is in a company's interest to see that its employees learn to love change. Automating the office is the only way to ensure much-needed productivity gains; automating the office must better the quality of working life or employees will see to it that those gains are chimerical.

The ultimate task of change managers and corporate educators is a very large one; it is nothing less than the establishment of a whole new mentality, the change mentality. The alterations necessitated by advanced technology in the office are so profound that it is not a question of adjusting just once but of living in a constant state of adjustment for quite some time. Adopting the change mentality will overcome many normal human fears, by removing the cause of those fears. Although living through this stage may not be easy, it will be both exciting and challenging. For those who have learned to love change, business, as it enters the age of information, is the place to be.

OFFICE AUTOMATION: PIONEER INDUSTRIES

Most analysts agree that how quickly office automation catches on depends on how much advantage over the competition it gives a company, rather than on how much time it will save a manager. It is not yet evident that manufacturers of primary products like steel or chemicals need to interchange electronic memos in order to remain competitive. For most primary industries the office of the future is a distant possibility, as yet a mere gleam in an ambitious young manager's eye. When the new technology is widespread and its contribution to office productivity an accepted fact, these industries will begin implementation. They will do so, benefitting from the mistakes of others, at their own measured and carefully planned pace.

Other industries, particularly those in the tertiary or service sector of the economy, do not have the luxury of time. The tertiary sector includes such industries as insurance, banking, travel, and investment, whose prime product is service. Electronic networking allows service industries almost instantaneous access to information. As a consequence, they can supply faster and better service to their customers. These industries exist on the cutting edge of competition: an extra service, a quicker decision, or a faster confirmation can make all the difference.

Service industries must rely on electronic information networks for a strategic advantage over competitors. Companies and even whole industries which do not implement the new technology fast enough will, quite simply, perish. For many of these industries, the changeover to the office of the future has already occurred. By necessity, the service industries have become its pioneers. The transformation of the airline and travel industries is an example familiar to the general public. In the face of rising costs and declining markets, these industries provide faster service and relatively low prices. Today's intricate fare structures and prepackaged plans give almost customized attention to every traveler as well as instant confirmation of all reservations. No human could hope to keep track of all the information required, but a computer can. And only a global electronic network instantaneously linking the entire world could hope to supply the information as rapidly and as accurately as needed.

In other service industries, automation is being introduced with amazing speed. Electronic communication is completely changing American business. It is no exaggeration to say that entire, well-established industries which don't go electronic quickly enough may be totally wiped out. Banking, a case in point, is well worth examining in greater detail. For our old familiar banks are on the verge of a metamorphosis. Indeed, if banks do not change their ways and do it faster than the competition, the entire industry may go under. Even banking, whose 1981 pretax profits of $21 billion make it one of America's giant industries, finds automation the only solution to its current problems.

The development of retail electronic banking skyrocketed in the last half of 1981. Within the next two years, experts predict that almost all banks will become affiliated with one of about six national electronic banking networks now springing up across the United States. Part of the reason for this unexpected spurt is that automated teller machines (ATMs), first introduced in 1970, have become cheaper and more reliable, prompting greater consumer acceptance. The installed ATM base nearly doubled in the past three years to 30,000 units. Bent on exploiting this potential, banks have entered into sharing agreements allowing ATMs of different banks and regions to accept the same debit cards. Bank computers link through leased lines and computerized switches. Computer service companies have also jumped into the fray, attracted by the electronic funds transfer (EFT) market, which may approach $2 billion by 1985.

But the major reason for this precipitous and costly stampede on the part of the normally conservative banks is simple: survival. Banks' problems began when money market funds started to attract people who had previously used bank savings accounts. Merrill Lynch's highly successful cash-management account allows investors to tap funds in their money market accounts by writing a check or using a VISA card. This innovative scheme was immediately copied by other brokers, whose parent companies have hundreds of retail outlets, linked by computers and supported by huge credit card bases. Federal regulations prevent United States banks from operating retail outlets outside their home states, but brokerage houses are not so restricted. So it seemed that brokerage houses could quickly grab control of the banks' traditional means of attracting depositors, and by operating nationally, even do the banks one better. Sensing catastrophe, the banks banded together to form nationwide EFT networks. It now appears that full-service electronic banks will be a countrywide fact as early as 1985 when deregulation of the banking industry will have been completed.

Automation is the only way for the banks to survive. For the past few years, attracting depositors' money has become increasingly difficult. To compensate, banks have been forced to increase services, such as offering interest on checking deposits or calculating interest daily for savings accounts. Improving services means higher labor costs, however. Consequently banks are seeing a lower return on their assets. Clearly banks must increase their revenues on transactional services that now generate little profit. The rub here, of course, is that costs on service have been running out of control. Individual banks alone have been singularly unsuccessful in electronic banking since they could not produce volume sufficient to cover the costs of the system. Now with national networks coming in, checks are on the way out.

The average cost of processing a check has jumped 36 percent in the past two years to forty-one cents; today the consumer usually pays only one fourth of that cost. Most people do not realize what a burden checks are for the banks. In New York City, for instance, each day clerks haul thousands and thousands of checks into the New York Clearing House. Checks, of course, are rarely deposited into their banks of origin, so the financial houses exchange their paper and settle up accounts several times a day. In New York City, this little transaction amounts to roughly $200 billion a day. Although the checks are physically delivered to the clearing house in huge boxes and bins, no money actually changes hands, for computers tabulate and exchange the amounts owing to different banks. It seems inconceivable that people are physically carrying sacks of checks on their backs, as it were, in this day of the mighty computer, but the fact is consumers like to see cancelled checks. The clearing house does also have an electronic system, known as the Clearing House Interbank Payments System, whose computer passes about a trillion dollars a week in checks, mainly from giant corporations.

Nationally the closest thing to EFT is the seven-year-old National Automated Clearing House Association, run by the Federal Reserve. For technical reasons, a typical transaction takes two days to complete, and as a result the system has replaced less than 1 percent of all check processing. To persuade customers to give up paper checks and switch to the electronic version, banks will use what is euphemistically called incentive pricing. By setting transaction charges at, say, twenty-five cents for checks, ten cents for automated tellers, and twelve cents for pay-by-phone, banks have already noticed a drop in the growth rate of check volume. Check volume grew by 7 percent annually throughout the 1970's yet, with incentive pricing, dropped to only 4 percent growth in 1981, and should level off by 1985 and decline sharply thereafter.

VISA and MasterCard, in conjunction with brokerage houses, have challenged the banks in the retail operations market. In addition to salvaging their consumer deposit share, banks also intend to use their electronic networks to openly compete with these credit card systems through electronic point of sale (POS) terminals. Most bank networks believe they can profitably provide POS service at a cost to the retailer of no more than forty cents per transaction as compared to the one dollar average credit card fee. And since the item purchased is paid for instantly, POS service will not only eliminate for the retailer the present two-day float period on goods purchased by check or credit card but also the clumsy telephone authorization method by which credit card companies were defrauded of an estimated $700 million in 1981.

ATM transactions are expected to rise from 1.5 billion in 1981 to 7 billion in 1985 and eventually to be dwarfed by POS and home banking operations, which could reach 8 billion transactions by 1986. Profits to the banks should increase. Each ATM machine costs about $30,000. To justify this expenditure, an ATM must run up an average of 5,000 transactions per month or the equivalent of the cost of a human teller. This occurred for the first time in 1981. When an ATM reaches 10,000 transactions per month, its cost is at least one third less than that of a human teller. By placing ATMs in supermarkets and other retail sites, banks can afford to close their brick-and-mortar branches while simultaneously increasing deposit-taking locations by 50 percent a month.

For some the loss of the familiar teller may be an occasion for sadness, even fear. It is clear, however, that banks have no choice in the matter. They must go electronic or the industry will die. That is far more threatening to the well-being of the economy as a whole and to the thousands of people who depend upon the banking industry for their livelihoods than the loss of tellers can ever be. To make up for this loss, the banking industry will spawn new and different occupations from its new electronic involvement.

Other pioneer service industries, already well launched into the "wired world" technology of the future, can show more precisely how electronic networking creates new jobs and transforms old ones. Most investment houses started forming electronic networks about ten years ago. At the beginning of the 1970's, successive waves of economic upsets hit financial houses hard. Inflation, for example, both domestic and foreign, altered the traditional behavior of markets; interest rates became extremely volatile; the formation of a solid OPEC cartel controlling world oil reserves and prices sent shock waves throughout the boardrooms of the world. For the first time, average investment houses, trading mainly on a national basis, were forced to the realization that the world had become a very small place. Events occurring in faraway countries would henceforth profoundly affect their markets and their profits. The obvious solution to the financial crisis lay in establishing electronic communication and information networks around the globe. In this way, investment houses could keep abreast of happenings in distant centers; through international trade, they could also influence those events. Once again, technology came to the rescue. Just as today's banks must automate or perish as an industry, investment houses of ten years ago had no choice in the matter. Maintaining solvency was the only justification for making the gigantic capital investment needed to establish the networks. And, in fact, many financial houses did not respond quickly enough: the 1970's was a period of massive corporate demise in this industry.

Full-service investment houses handle diverse financial transactions: estate and tax planning, stock purchase, and securities analysis, to name a few. Automation has made many changes in these operations and services. Electronic networks not only enhance the quantity and quality of services by supplying vast amounts of information instantaneously from across the globe but also permit financial houses to trade internationally. Because of the networks, the world is now one interconnected financial unit.

One new service which barely existed ten years ago has caught the imagination of the public: the money market. In essence, the money market is a huge reservoir of funds, made up of contributions from many would-be lenders, available to borrowers, at a price, on a short-term basis. Although the principals are usually multinational corporations, banks, and governments, recently, by pooling their resources, even middle-income earners have been able to join in. The object of the money market game is for both borrowers and lenders to maximize their profits: lenders by securing the highest possible rate for the use of their capital and borrowers by placing the capital where it will gain the most. It is a game of split-second timing and gut-wrenching decisions. In its present form, the money market has existed for about fifteen years; mechanisms permitting ordinary individuals to contribute small amounts of money have been in existence for about ten years in Europe and seven years in North America.

Although theoretically the short-term borrowers and lenders could find each other and establish their own terms of cooperation, practically such a union would rarely take place. So a new occupation came into being: the money market trader. Money market traders, who generally work for large investment houses, are the go-betweens who put the lenders and the borrowers into contact and set the terms. Bob Lavers works in Toronto as a money market trader for Midland-Doherty, one of Canada's largest stockbroking firms and the third largest investment dealer in the country. At thirty-eight, as one of the more experienced traders in his incredibly hectic business, Bob is in a good position to comment on the changes wrought by automation. Surrounded by digital stock readouts, a bevy of blinking computer monitors, ringing telephones, and people yelling at the top of their voices, Bob keeps one eye on the screens and one ear on the telephones as he calmly explains why his job exists.

Midland-Doherty and other houses have worldwide financial information networks which gives them the technical capability of putting borrowers and lenders together almost simultaneously. To make their decisions, traders need to receive and correlate prodigious amounts of information, usually at the same time. Bob follows markets all around the world: gold in London, wheat in Chicago, cotton in Louisiana, Eurodollars in Singapore, silver futures in New York, the Mexican peso in Montréal, and many more. Information floods in via the new electronic capability at the traders' disposal. Seated at a console with ten other money market traders in a room crammed full of other market groupings, Bob screams through the din that Midland-Doherty's seven-year-old equipment is already obsolescent. In front of every trader is an endlessly jangling ninety-button telephone set with direct lines to many competitors; shared by all at the console are three computer monitors constantly screening new market updates, an intercom via land-links to the Montréal office, and a receiver for the satellite hookups from thirty or forty spots around the world. Information, written and spoken, rolls in ceaselessly throughout the working day. To the visitor, it means instant headache; to the trader, it means remuneration in excess of $100,000 a year.

Electronic networks also bring speed. On the money market, speed is everything. Traders rely absolutely on the information screened by the computer; it must be instantaneous. Before the screens, information lag time averaged five minutes. Today, when information lags by even a few seconds, traders find out about it the hard way, by losing money. Because of a three-second delay, this, in fact, happened at Midland-Doherty. The discount rate of a certain bond had dropped by .5 percent, but due to a technical snafu resulting in the three-second screening lag, the traders did not know about it. During those three seconds, competitors sold them bonds at the old rate. For want of three seconds, the company lost millions of dollars. As the lifeblood of the industry, the information networks must be technically perfect, their speed far beyond what mere mortals could provide.

In investment houses, computers are essential for three tasks. Computers provide the inventory control over the millions of dollars in securities held routinely by larger houses. They screen and correlate the information received from other markets. And they can help in the complex technical analysis of market trends. Analysis is really the soft underbelly of the whole trading business. Traders must not only have the mental capacity to assimilate great quantities of information, they must also make snap decisions based at the moment primarily on informed intuition. Since a large company does upwards of several billion dollars of business in a week, the pressure is intense. Traders know that when dealing with sums of that magnitude, one serious error in judgment could close the doors of the firm forever. The decisions arise from multilayered, labyrinthine mental processes. To take a simple example, traders must decide what the latest unemployment figures may mean to the market. If the figure rises from 9 percent one month to 10 percent the next, this is good news for the holders of interest-bearing securities. By a curious process of double-think, though, it may be bad news, if expectations had been for a jump to 10.5 percent. Traders must understand such seemingly extraneous facts and much more. The news of the day causes wild stock fluctuations; traders must be well informed. A computer can help in this difficult task of market analysis. But that old bugaboo, software, is still a major hindrance.

Many compare the job of a money market trader to that of an air traffic controller. Traders, like controllers, assimilate great chunks of information from diverse sources simultaneously. Continuously making important decisions with no time for reflection, though, engenders stress: the burnout rate is high; the average age of a trader is thirty-three. Alcoholism and divorce rates are also higher than normal. Unlike air traffic controllers, traders do not need any specialized training. Traders come, not from a certain background, but rather with certain personality characteristics. To be good at selling, a trader needs to be aggressive, with a strong ego and endless self-confidence. Recent university graduates display a notable lack of drive, according to working traders. "You have to light a fire under them," they grumble, and no one has the time. Other desirable characteristics include a good memory, the rare ability to both talk and listen at the same time, and general equanimity of mind. Traders cannot be moody. Traders must also be articulate and skilled on the telephone; almost all business is conducted by phone.

Some of the traders at Midland-Doherty come from the London School of Economics, some are accountants, some are MBAs. The best trader in Toronto is an eleventh-grade graduate who came up as a messenger boy. And traders prophesy that one twenty-one-year-old woman who started as a secretary will be on the Board of Trade before her thirtieth birthday. Although the pressures of this job are great, so are its rewards. One of the most attractive aspects of this job, aside from a high income potential, is the instant gratification it provides. Traders know instantly when they've made the right decision.

Though without advanced technology the job would not exist, no machine could take the place of a money market trader. As in other industries, technological advance creates, as it destroys, jobs. Information brought by the new electronic capability, now at high-tide level, threatens to engulf the puny human mind in a gigantic tidal wave. Yet, experts say, current levels of information are insufficient. There are, indeed, no perceived limits to its growth. Only human labor can originate, comprehend, sift, correlate, and finally act upon information. Thousands of jobs are in the making here in this tiny part of the coming Age of Information.

AFTER AUTOMATION: AN ECONOMY OF THE FOURTH LEVEL

The appearance of the microprocessor has ushered in a revolutionary era. As yet, we can catch but a glimpse of the new Age of Information and have but a small idea of its enormous social consequences. The automated office, along with microcomputers, videotex, and all the other innovations considered in this book, is a transformative technology. Implementing the office of the future will have repercussions far beyond the office. The many sophisticated technologies of the integrated office will transform our lives, just as the watershed inventions of the Industrial Revolution irredeemably altered the lives of our ancestors and so the course of history.

One of the many seminal inventions made near the beginning of the Industrial Revolution was the steam engine; the steamboat was one of its most revolutionary applications. A look at the many changes the steamboat brought about can help us understand the stunning impact the new automated office will have on the workplace of the future. The primary purpose of the steamboat and the automated office is to provide a technical solution to a practical problem. Both, however, have unforeseen, secondary consequences which are of far greater import, for both are transformative technologies. Just as the invention of the steam engine assured the success of the Industrial Revolution, the converging technologies of the office of the future fling open the doors to the coming Information Revolution. The Information Revolution, in a way we are just beginning to comprehend, not only puts a definitive end to the Industrial Revolution but in a queer paradox functions as its logical extension.

In its time the steamboat addressed a major problem of the emerging Industrial Revolution: how to get manufactured goods to expanding markets quickly and cheaply. This machine and its successors solved the problem more than adequately. A sailing ship could go around the world in two years; a steamship took a mere two months. For the first time in human history, a dependable and amazingly rapid global transportation network became a reality; people and goods could travel freely and quickly throughout the entire world. But the steamship did something else. Although conceived of as a transportation device, the steamship incidentally also eliminated the stubborn obstacles to communication plaguing every civilization throughout recorded time. With the incredible speed of the new machines, the temporal and spatial limitations on communication and transportation, previously thought intrinsic to nature itself, suddenly melted away forever. In the past, the far-flung empires of countless civilizations had foundered on the barrier of distance; distance renders communication difficult and thus political control almost impossible. Increasing the speed of transportation and so communication, two sides of the same coin, now became a perceived goal of the developing Industrial Revolution, and eventually its most durable achievement. What had begun as a simple, practical solution to a purely mercantile problem evolved into a solution to a far more fundamental difficulty.

The steamboat transformed the physical world and the everyday lives of succeeding generations. Clearly, among its descendants must be numbered the automobile and the airplane, our interstate highway system and global flight paths. These material changes in our environment resulting from the introduction of the steam engine are truly remarkable. Transformative technological innovation, however, also alters modes of thought and perception: here lies its most profound effect. The world had never seemed so small a place in all the long history of the human race as it did after the appearance of the steamboat. Seeing how rapid and extensive the material changes were, people's expectations rose. Progress seemed possible, even inevitable. For the first time, the very stars beckoned. What had previously been beyond dreams now seemed thinkable.

As the steamship with its unimaginable speed cracked the intransigent barriers of time and distance, it thus, for a time, linked communication and transportation inextricably. Transportation, however, despite successive technological leaps, remained tied to the material in the form of its cargo—people and goods—and thus subject to physical limitations. Communication was not so hindered and shortly took off on another track with the invention of a range of modern devices, the telegraph, the telephone, radio, and television, new conquerors of distance. Collateral descendants of the steamship, these direct progenitors of today's Information Revolution diminished for communication the importance of physical space almost to the vanishing point.

Like the steamship, the office of the future came into being as a solution to a practical problem: in this case, declining office productivity. But like the steamship, the automated office is a transformative technology with ramifications far beyond its primary task. As the heirs to two hundred years of technological innovation, we have learned to look for its consequences. Already we can dimly see how the office of the future will alter our physical world and, more fundamentally, our traditional ways of ordering experience.

Establishing the global information networks ultimately necessary to the office of the future will initiate the total collapse of all time and space limitations for transportation as well as communication. For the converging technologies of the office of the future replace the durable goods of the moribund manufacturing sector with a new product—information. Information is non-physical like voice; information can be transported nearly instantaneously over global networks. The place of origin or the distance between transmission points means nothing: data transfer from Hong Kong to Cape Town takes no longer than between Manhattan and the Bronx. So with the coming of the automated office, communication and transportation are once again indissolubly linked.

Replacing manufactured goods, information will be a major form of wealth in the postindustrial world of tomorrow. Yesterday's world, where most of us still live, was brought into existence by the Industrial Revolution; it displays a three-tiered economy. The primary sector manufactures basic products like cars; the secondary sector manufactures goods to support the first, such as car parts or tires. The tertiary sector supports the whole economy by providing services that range from dry cleaning to social work. As sources of employment, the primary and secondary sectors of the American economy have been declining in importance for the last thirty years. Growing at a rapid clip in North America since the

1950's, the service sector has now nearly reached its limit in size. Already futurologists have dubbed the newest sector, created by information technologies, the quaternary, or fourth, level of the economy. The fourth level will expand exponentially throughout the rest of the century. It will be the prime job-creating sector. It will completely alter traditional notions of production and wealth.

For us, born into the last days of industrial society, an economy not founded on a physical product, like potatoes or toasters, may be hard to imagine. Yet already the new economy replacing manufacturing surrounds us. The New York garment district provides a most striking example of the changeover to an economy of the fourth level. The creators of status jeans, such as Jordache, Sasson, Bon Jour, and Murjani, do not manufacture the product; manufacturing takes place in Hong Kong, South Korea, and China, where wages are low and unskilled labor is abundant. New York, however, retains what has been called "the knowledge shell," a cadre of management and clerical staff linked to support groups of lawyers, accountants, and media specialists who merchandise the jeans with high-voltage advertising. For both seller and buyer, the information exchange makes up 90 percent of the entire transaction. The product is really of negligible importance. What counts is promoting a name and an image. In fact, the wealth created by the New York garment district now comes no longer from producing a tangible product but mainly from information manipulation.

The Port of New York and New Jersey Authority, a gigantic umbrella agency, supplies another example of how old concepts are transformed into new industries. Recently officials of the Port Authority have become aware that its main business is not the construction and maintenance of bridges, roads, tunnels, and piers, for which the authority was founded. Suddenly it is obvious that the business of the Port Authority is information. The World Trade Center, within its domain, houses companies engaged in national and international trade. But the center's 4 commodity exchanges, 90 steamship lines, and grand total of 900 companies can barely keep up with the information load, not to mention ahead of the competition. In response to the need, the Port Authority envisages a new communications center to be called a teleport. In partnership with Comsat General Corporation, the

Port Authority will set up a central office to receive and distribute messages from its own orbiting satellites. The teleport could, within a few years, make the Port Authority's revenues from land and waterborne commerce look anemic.

Among the many diversified elements making up the fourth level of the economy are, of course, all its electronics-based technologies, many of them discussed in this book. Computers, satellites, videotex, telecommunications, and automated office equipment, fiber optics, and lasers are just a few of the many. Other technical elements might include software and data base creation. Surprisingly perhaps, these highly visible and much-publicized new manufacturing enterprises are probably the least important part of this sector for long-term job creation. Certainly, over the short term, manufacturers, rushing to supply an escalating demand for information technology, will also create jobs. Over the long term, though, it is virtually certain that automation in the form of industrial robots will take over by-now-routinized manufacturing. Alternatively manufacturing may be done in countries where wages are lower, much as now happens in established technologies like television, radio, and stereo manufacture. In the economy evolving after the Industrial Revolution, the manufacturing sector took pride of place, supplying the bulk of employment.[1] In the postindustrial world of tomorrow, manufacturing will play a much smaller role. Thus establishing computer factories to replace assembly-line jobs lost in traditional manufacturing is not necessarily a step into the future or any guarantee of a continuing source of employment. If manufacturing is the sole element of the information sector to be implanted in a given locale to generate employment, it is an example of delusional thinking, based on an outdated model of economic activity. Relative to that of underdeveloped nations, the work force of developed countries is literate, skilled, cultured, and therefore flexible. For the most part, employing such a populace on assembly lines is a serious misuse of a unique resource. These people should be producing information.

The term information, as it is applied to the fourth level of the economy, means the data produced by computers, "number crunching," but information is also fact, opinion, persuasion, and decision. Seeing it in this light, one can more easily understand why the fourth level takes so many forms. Any idea for collecting, arranging, packaging, or distributing information, in whatever form, can further expand this sector, which is already the fastest growing in the economy. Humanity has a perpetual hunger for knowledge and intellectual innovation. There are, therefore, no perceived limits to the growth of information. Providing and disseminating information in its many forms is the purpose of the new technologies; using the information is the concern of the greater part of the fourth sector of the economy. The non-manufacturing components of the Information Society have almost limitless potential for growth and for job creation. Information markets, unlike those for manufactured products, can never be saturated.

The large urban center, previously the site of the most intensive manufacturing activity, is now the focus of greatest growth in the information sector. Cities have, of course, always functioned as information and communication centers, but the old structures and institutions are now being transformed by technology. A few examples provide graphic illustrations of the tremendous changes which have already occurred.

If information is the currency of the future, then the humble library, transformed by electronics, is the bank of tomorrow, containing untold wealth within its invisible walls. A city generally has good libraries. The many tasks associated with creating, organizing, coding, filing, and distributing the information currency held by these banks will provide employment throughout the 1990's. Anyone who has not walked into a public library in the past few years is in for a surprise. Librarians are no longer little old ladies whispering "hush!" and stamping out books but rather trained "information systems specialists." The library still has books, but it also has videotex, data bases, and computer-assisted research facilities. The University of Toronto's library system, for example, with over 5 million items, is the largest library in Canada and a primary information data base. Its cataloguing system, UTLAS (University of Toronto Library Automation Systems), the second largest in the world, is a data base containing more than 5.6 million entries. It connects to other systems as far away as Japan to form a massive international cataloguing service. The little old librarian of not so long ago would faint dead away.

Of course, it isn't the university, guardian of civilization, which will turn information into money. Rather, barbarians like Toronto's Infomart and its rivals, InfoGlobe, Micromedia, and QL Systems, will loot the university's primary data base for information they will use for commercial videotex data bases. All information entrepreneurs will know the library well.

Yet once again, it can be seen how information technology diminishes the importance of distance and, not incidentally, of national boundaries. It is worth remarking that computerized libraries with gigantic data bases add new significance to the term information. Data absorbed into the individual's mind we call knowledge. The contents of an unread book is information available to a potential reader. Computerized data bases, however, store data in carefully structured ways so that retrieval systems can operate rapidly and effectively. Such information, while not as accessible as personal knowledge, is far more available than data stored in a book and buried somewhere in the library. Occupying a middle ground between information and personal knowledge, such data can be eminently accessible to the ordinary citizen. Keeping in mind the old adage "Knowledge is power," who knows what consequences that may have for the future?

The gigantic data bases of tomorrow's library-banks must get their information-currency from somewhere. Truly new information can come only from the human mind. A city is also a cultural and educational center, a kind of intellectual crossroads. Inventive, imaginative, and productive people meet and mingle in the city, stimulated by new points of view and ways of thinking to create yet more ideas and new ways of thinking. Such people are the information creators and so a priceless resource for the fourth level. They supply both the raw material and the finished product. They collect, they arrange, they package and distribute their creations. They have opinions or make opinion; they persuade; they decide. Without their efforts, the Information Society would collapse in a moment.

In Toronto's city center, for example, there are forty theaters, three television networks, several radio stations, a dozen book publishers, three newspapers, four institutions of higher learning, and six yellow pages' worth of advertising firms. Civil servants, computer specialists, symphony soloists, seven thousand lawyers, professors, dancers, consultants, actors, media specialists, theatrical agents, film makers and producers, sixty-eight hundred chartered accountants, and countless doctors and hospital staffers are just a few examples of those engaged in information-related occupations. Toronto, though the hub of Canada's cultural life, is a city of less than three million people. New York City, one of several major American information centers, personifies information on a gargantuan scale. In fact, estimators of New York City's gross city product, the total value of goods and services produced, think that information-related activities generated more than half the 1980 $100 billion total. The variety of occupations created or transformed by the new information technology can hardly be imagined.

Tracking information creators is an endless task, but naming a few of the many new or altered occupations can give a faint idea of the immensity of the creative possibilities offered by the new technologies. Computer market analysis allows magazine publishers to tailor a publication so exactly to a particular readership that specialized audiences receive greater support than before. Cabaret acts, rock groups, and other performers, once confined to small clubs or giant, acoustically disastrous stadia, can now cut video disks of a performance and play to a much wider public. Nightclubs in many cities, equipped with the latest technological innovations, project performances recorded on the disks, which have exceptional acoustic fidelity, on giant screens and present local live entertainment during the same evening. By maintaining its position at the cutting edge of technology, the record business can also flourish in the new information sector. Performances by world-class symphony orchestras before live audiences can reach millions through cable. Museums, like libraries, are treasure troves of information, before available only to the privileged few. By using the combined expertise of many information creators, world-famous museums, such as the Smithsonian Institution or the American Museum of Natural History, can convert dusty exhibits into data bases for use in classrooms, on television programs, or as accessible research material. Through astute application of the new technologies, museums can make their most prized acquisitions the intellectual property of millions who would never otherwise have had the opportunity.

Under the influence of the new technologies, even the professions are changing. Information technology allows accountants and lawyers to make it through tax time with fewer migraines and twenty-four-hour working days. As a result, such professionals can spend more time researching better deals for their clients, who can be anywhere on this globe. New technology makes the research faster, too. Educators, of course, are also affected as instruction, especially in the sciences, is increasingly computer-assisted. Residents of remote communities now regularly receive classroom instruction via satellites. Once the technology is in place, handing in homework or asking questions electronically could make such education more personal. Some even foresee that the role of the teacher will change to that of instructional manager. Medicine and treatment are now information-based. Continent-wide data bases facilitate organ transplants through a retrieval and exchange program. Computers now even aid in diagnosis. Specialized diagnostic computer programs are a special aid in teaching diagnostic skills to the inexperienced. Many metropolitan areas use computers to match distress calls with the nearest available ambulance. Technology, in fact, is what distinguishes American medical treatment from that of the rest of the world. Cities are eager to encourage high-tech medical centers. In Miami, for example, well over a billion dollars a year is generated in medical fees and supplies; its top-flight medical facilities and renowned physicians attract many, especially wealthy Latin Americans, to the city.

A city is a financial center as well; business and finance have been changed beyond recognition by new technologies. During the big "paper crunch" of 1969–70, the New York Stock Exchange nearly folded under the pressure of trading 11 million shares a day. Since then, computerization has vastly expanded its capacity. Today, 70-million-share days are routine. In the near future, the Big Board will easily handle 150 million shares per day. When videotex is widespread, stock buyers will be able to call up their own research and execute their own orders with their smart cards. Firms whose main business has traditionally been in the realm of finance are now dependent upon communication technologies. American Express, for example, a leading force in consumer credit, is a part owner of Warner-Amex, a cable company. This complex and growing interrelationship between communication and finance will create many employment opportunities; especially those job seekers with the broad-based skills of a general education will have sufficient flexibility to handle these jobs.

The information sector, like the service sector, continues the trend toward increasing levels of abstraction. Its product is not physical. And the technology of the automated office, for example, transforms one communication medium into another with amazing ease and speed. Business letters, files, reports, and even money need no longer materially exist; their symbolic representation as digitized squiggles, stored somewhere on computer-compatible material, is sufficient. Since a letter is already an abstract representation of speech, a symbol for actual language, the leap to digitized representation of letters adds yet another layer of abstraction. Similarly money is already a symbol for human labor; representing it as an electronic bleep is another layer of abstraction. Without doubt, the expansion of the fourth level means increasing intellectualization of reality. Where it will bring us, nobody knows. Yet one thing is sure: The new information technologies will require an educated, cultured, and extremely flexible work force. The only limitations to the expansion of the new technology are the limitations on human creativity.

INDEX

LIBRARY
ST. LOUIS COMMUNITY COLLEGE
AT FLORISSANT VALLEY